Conflict Resolution in Coastal Zone Management

Environmental Education, Communication and Sustainability

edited by Walter Leal Filho

Vol. 28

PETER LANG

Frankfurt am Main · Berlin · Bern · Bruxelles · New York · Oxford · Wien

Walter Leal Filho
Nils Brandt
Dörte Krahn
Ronald Wennersten
(eds.)

Conflict Resolution in Coastal Zone Management

PETER LANG
Internationaler Verlag der Wissenschaften

Bibliographic Information published by the Deutsche Nationalbibliothek
The Deutsche Nationalbibliothek lists this publication in the Deutsche Nationalbibliografie; detailed bibliographic data is available in the internet at <http://www.d-nb.de>.

Copy editing and Layout:
Kumpernatz + Bromann
www.Kumpernatz-bromann.de

ISSN 1434-3819
ISBN 978-3-631-57375-4

www.peterlang.de

Summary

Table of Contents

Introduction

Coastal zones are areas where the land meets the sea. The definition of coastal zones therefore includes both coastal waters and adjacent shorelands. Coastal zones have a conspicuous structural diversity, which includes beaches, dunes, cliffs, estuaries, lagoons and others features such as mangrove swamps. Coastal areas are important habitats for a wide number of plant and animal species.

Due to their local and geomorphological structure, coastal zones are particularly vulnerable to pressures from natural processes such as shoreline erosion or biodiversity losses, as well as pressures from mankind such as construction/building developments, economic activities such as tourism and fisheries, as well as industrial production, which may generate non-point source pollution.

In view of the importance of coastal areas to Baltic countries, integrated coastal zone management (ICZM) in the Baltic Sea Region is a matter of prime concern. The percentage of the population living within 100 km of the coast is 100% in Denmark and 88% in Sweden. In other Baltic countries too, a substantial proportion of the population lives in or close to coastal areas. The aim of ICZM in the Baltic is thus to protect the structure and the fragile natural resources that exist in coastal areas by reducing conflicts between competing land and water uses, while representing a comprehensive approach to managing the impacts of development and other activities on coastal areas.

ICZM is an important feature of the project *Coastal Zone Management in the Baltic Sea Region* (COASTMAN), undertaken in the context of the Interreg IIIB (Baltic Sea) Programme. The main objectives of the project are to:

- Demonstrate the conditions under which conflict resolution related to coastal zone management can be handled in a bottom-up perspective, starting from practical case studies covering the multiple dimensions of the problems
- Promote long-term capacity building in ICZM, with the focus on methods for conflict resolution through the development of a network and joint educational programme for coastal zone management in the Baltic Sea Region.

The rationale for the COASTMAN project was that although coastal zones in the Baltic region have a great potential for economic growth, there are several serious threats to ICZM in the region. Two of the main threats to sustainable development are conflicting interests and goals, and environmental pollution. Potential conflicts are multidimensional and include cultural and spatial dimensions, while environmental threats are also very complex and wide-ranging. Therefore the COASTMAN project tried to look at the problems and at concrete approaches to tackle them through national case studies in the participating countries. This approach was used in order to maximise the potential for exchanges of 'hands-on' experience between the different countries taking part in the project and at the same time create a mutual learning experience regarding conflict resolution strategies and methods in the Baltic Sea region.

The experiences obtained as part of COASTMAN are collated and presented in this Handbook, which begins with an overview of the issue of conflict resolution in coastal zone management. This is followed by description of a set of administrative and legal processes related to coastal zone management and by an overview of the methods that can be used in conflict resolution in coastal zone management. The Handbook then continues with a set of case studies in the following regions:

- Hamburg (Germany)
- Haapsalu (Estonia)
- Klaipėda (Lithuania)
- Primorsk (Russia)
- Stockholm (Sweden)
- Ventspils (Latvia)

In addition, a case study outside the Baltic Sea region – Guanabara Bay in Brazil – is included in order to illustrate the process of framework analysis and its usefulness in understanding environmental conflicts.

The next part of the Handbook comprises what can be referred to as the 'education' section, in which information relating to the educational and training aspects of conflict resolution in ICZM is presented and many relevant aspects for consideration are discussed. In particular, this section describes a pilot course on conflict resolution in the Baltic Sea region, which provides a deeper knowledge and understanding of conflict resolution and the use of different methods, tools and experiences on conflict resolution in the region.

The last section of this Handbook consists of the conclusions, where the main lessons learned are presented and future perspectives are outlined. Thanks to its scope and trans-national dimension, this Handbook will be instrumental for organisations responsible for ICZM and will provide some inspiration towards initiatives that can be undertaken at the local level in order to better understand, and hopefully address, conflicts arising in coastal areas.

Walter Leal Filho,
Nils Brandt,
Dörte Krahn,
Ronald Wennersten

Summer 2007

Chapter 1

Administrative and Legal Processes in Coastal Zone Management

Ronald Wennersten[1]

1. Background

Cities in coastal areas, and on islands in particular, have a special interest due to the role they may acquire as growth poles and gates to their hinterland in a broader process of spatial integration but also as centres of growing economic activities such as tourism, transport and fishing. Furthermore, they are important as areas facing increasing pressures from a multitude of development actors with diverse interests. As a consequence, their planning should be seen in the context of broader spatial strategies (1).

There are various definitions of a 'coastal zone' and Coastal Zone Management. In order to understand the administrative and legal processes in Coastal Zone Management in the Baltic Sea Region, knowledge about the various definitions applied to coastal zones is fundamental. These definitions can be found in Appendix I and II.

In the future, both population growth and economic development worldwide will continue to be concentrated in coastal regions. In Europe, high growth is forecast for the Baltic Sea Region in particular, and its coastal zones will play a central role. These will be subjected to constantly increasing pressure from human use, which can create conflicts between different stakeholder groups (2).

The increasing pressure on the coastal zone through anthropogenic activities, such as building of new roads, harbours, marinas, hotels, private summer houses and camping grounds, poses a permanent threat to its biological, geological and geomorphological values, as well as to sustainable use of the coastal zone (3). In many countries, landward port expansion is becoming more difficult as the established ports are traditionally located near the city centre (4).

In the terms of the EU Water Framework Directive, the Baltic is designated a coastal sea and an 'ecoregion' to be considered as an integral whole. It represents a model region for the development of 'marine awareness' within the Integrated Coastal Zone Management (ICZM) framework (2).

Coastal zones around the Baltic Sea are of great importance to many people in the surrounding nations, as these zones have high population densities, hold a

1 Royal Institute of Technology (KTH) – Department of Industrial Ecology – Stockholm – Sweden.

large proportion of available work and development opportunities and have an extensive transport network. However, the intense use of coastal areas in the Baltic Sea region has brought about severe environmental problems and the environmental situation has continued to deteriorate rapidly in many parts of the Baltic Sea.

Since the activities within Baltic Sea region are interlinked and likely to affect another, it is important that problems associated with the use and development of zones around the Baltic Sea are solved in a flexible way involving all neighbouring countries.

Furthermore, the environmental problems within coastal zones of the Baltic Sea call for international action. The necessity of international cooperation is due both to the character of existing problems and to the fact that the resources of the Baltic Sea are shared between member countries of the European Union (EU), including Sweden, Finland, Estonia, Latvia, Lithuania, Poland, Germany and Denmark, and the non-EU member country the Russian Federation. Measures taken at national level in each country in order to solve the current environmental problems of the Baltic Sea region have proved to be inadequate. Pollution of the Baltic Sea is spread across country borders, implying that cooperation is necessary between all neighbouring countries, including both EU member and non-member states.

There are a number of generally accepted principles and characteristics in the implementation of Integrated Coastal Zone Management. One of these is legal support, which allows application of conventional norms of international law in the regulative mechanism of coastal management.

The role of the law in realisation of the concept of *integrated management* is that it defines the powers and duties of many state and private organisations and also natural scientists involved in management and use of the coastal zone, providing a legislative basis for the framework in which they function. Hence, the law has the potential to encourage the ICZM process, but at the same time can be the factor constraining this process.

A legal, administrative and regulatory basis is necessary for taking decisions and for their implementation. There should be mechanisms that guarantee coordination in decision-making. Successful implementation of Integrated Coastal Zone Management is only possible if the legislative base is present, and where the coast is regarded as an integral natural object. Conservation of the natural resources of coastal territories, where the interests of various nature users clash, is impossible without an effective legal mechanism for the harmonisation of relations between the various users of the natural resources.

In general, international experience shows that artificially created legal distinctions between territorial and marine laws hinder implementation of ICZM. Coasts must be considered a complex area of legislative expertise.

The choice of legislative mechanisms for ICZM depends on the national legal system, economic conditions, the natural features of the coasts and the state of their environment. At the same time, national coastal legislation should accord with the principles and standards of international law. Despite the variety of laws that govern coastal zones in different countries, legislative problems associated with the coastal zone are remarkably similar worldwide (5).

The European Union's coastal regions are under constant pressure. Almost half of the EU population now lives within 50 kilometres of the sea (1) and coastal zone resources produce much of the EU's economic wealth. The fishing, shipping and tourism industries all compete for vital space along Europe's estimated 89 000 kilometres of coastline, and coastal zones contain some of Europe's most fragile and valuable natural habitats. However, the increasing demand for coastal resources is leading to their degradation – reduced water quality and quantity, accelerated erosion, accumulation of pollution, loss of fisheries resources, etc. Moreover, this degradation has negative social and economic consequences.

Europe's coastal zones probably face a larger number of economic, social and environmental problems than any other area of the European Union. From Lapland to Crete, coastal zones are facing serious planning and management challenges (6).

Different policies have been proposed to implement ICZM, but there is still a lack of a broader framework that promotes an integrated management perspective for coastal zones. Some barriers to ICZM realisation still exist, such as:

- Bureaucratic inertia
- Opposition to change
- Conflicts arising from multiple interests
- Overlapping of local, regional and national expertise
- Difficulties in cooperation between different bodies (1)

Bringing together all the local, regional, national and European policy-makers and other stakeholders whose activities affect coastal regions is central to ICZM. Without coordination at all levels, efforts to protect the EU coastlines will only have limited success. The stakeholders involved should include not only government officials and policy-makers but also other interested parties such as local residents, non-governmental organisations and businesses (6).

There are a number of different tools (carrying capacity, strategic environmental impact assessment, sensitivity mapping) that can assist in ICZM implementation. The planning process is one of these. It is an important and useful tool for ICZM, but in order to be effective it should be: 1) accompanied by a variety of policy instruments; 2) put in timely action; 3) developed in a strategic manner with a long-term vision and; 4) take into consideration the interconnections across spatial levels (1).

The Baltic Sea region is one of the most vulnerable areas in Europe and faces a range of interrelated biophysical and human problems. Some of the most common manifestations of these problems are:

- Widespread coastal erosion, often exacerbated by inappropriate human infrastructure and development too close to the shoreline
- Habitat destruction as a result of poor planning, poor land development or sea exploitation
- Loss of biodiversity, including decline of coastal and offshore fish stocks as a result of damage to coastal spawning grounds
- Contamination of soil and water resources, as a result of the migration of pollutants from marine and on-land sources to coastal zones
- Deterioration of water quality and quantity as water demand exceeds supply or wastewater treatment capacity
- Unemployment and social instability resulting from the decline in traditional and environmentally compatible sectors, such as coastal fisheries
- Competition between users for resources
- Destruction of cultural heritage and dilution of the social fabric following uncontrolled development – especially of tourism
- Marginalisation and emigration, compounded by a lack of appropriate infrastructure, including year-round communications and transport networks

The intensive and increasing anthropogenic pressure on the sensitive and vulnerable ecosystems and biodiversity of the Baltic Sea region means that Integrated Coastal Zone Management needs to be systematically implemented throughout the region.

The Baltic Sea is a relatively shallow inland sea surrounded by Denmark, Estonia, Finland, Germany, Latvia, Lithuania, Poland, Russia and Sweden. It also receives surface water drainage from several other countries. It is one of the major brackish water basins of the world and is physically dominated by the freshwater input by rivers and precipitation and limited inflow of more saline water via the shallow entrances to the North Sea.

Nearly 85 million people live in the Baltic catchment area, 26% of them in large metropolitan areas, 45% in smaller urban areas and 29% in rural areas. Population growth, urbanisation, agricultural intensification and land use changes have led to an increase in the use of fertilisers, increased industrial emissions and inadequate treatment of sewage.

All Baltic Sea countries are, without exception, engaged in ICZM work, although there is no specific legislation pertaining to ICZM in the Baltic region, which must use existing policies and instruments.

The EU has called for a very broad set of principles to be adopted concerning good coastal management. These are:

- A broad holistic perspective
- A long-term perspective taking the precautionary principle into account
- Adaptive management
- Local specificity
- Working with natural processes
- Participatory planning
- Support and involvement of all relevant administrative bodies
- Use of a combination of instruments (7)

2. Legislation affecting coastal areas in the Baltic Sea

There are several multilateral treaties that are of relevance in the Baltic Sea area, With a brief description of the most important given below. A more detailed description is given in the Helsinki Convention, EIA and WFD.

2.1 International agreements

2.1.1 United Nations Convention on the Law of the Sea (UNCLOS)(8)

The United Nations Convention on the Law of the Sea was adopted in 1982 and addresses *inter alia* marine environmental pollution. Its intention is to establish

> „a legal order for the seas and oceans which will facilitate international communication, and will promote the peaceful use of the seas and oceans, the equitable and efficient utilisation of their resources, the conservation of their living resources, and the study, protection and preservation of the marine environment".

UNCLOS calls for states to prevent, reduce or control marine pollution and safeguard and manage marine living resources.

2.1.2 Montreal Guidelines on the Protection of the Environment against Pollution from Land-based Sources

These guidelines are only recommendations, but they function as a framework for the development of regional agreements and perhaps a global convention. They give additional support to Article 207 of the United Nations Convention on the Law of the Sea, which requires states to '*prevent, reduce and control pollution of the marine environment from land-based sources, including rivers, estuaries, pipelines and outfall structures*'. The guidelines were developed from shared fundamental principles and standards of existing arrangements, as well as practical experience from their development and operation. They call for states to protect and preserve the marine environment. They also call for states to adopt measures to prevent, reduce and control pollution from land-based sources. The guidelines also suggest that states develop protected areas.

2.1.3 *International Convention on the Prevention of Pollution from Ships (MARPOL)* (9)

The International Convention on the Prevention of Pollution from Ships, or the MARPOL Convention, was adopted in 1973 and modified in 1978. This convention aims to eliminate all pollution of the sea by chemicals, oil and other destructive materials. It also aims to reduce the amount of oil that is released accidentally by shipping accidents. Moreover it aims to improve prevention and control of marine pollution from ships.

2.1.4 *Convention on the Prevention of Marine Pollution by Dumping of Wastes and Other Matter (1972 – London Convention)*

The London Convention (1972) encompasses all global seas, as well as the territorial waters of coastal states. It aims to prevent all arbitrary disposal at sea of wastes that could cause harm to human health, to living resources and to marine life. The main principle of the convention is to prevent all dumping of particular wastes, and to legislate the need for a permit for the dumping of other wastes.

2.1.5 *International Convention for the Safety of Life at Sea (SOLAS 1974/78)*

The SOLAS Convention covers the safety of life, the safety and efficiency of navigation and/or the protection of the marine environment.

2.1.6 *Convention on Biological Diversity*

The Convention on Biological Diversity from 1992 entered into force on 29 December 1993 following 30 ratifications. The convention aims to conserve biological diversity and to promote the fair and equitable sharing of the benefits from its use.

2.1.7 *Basel Convention on the Control of Transboundary Movements of Hazardous Wastes and their Disposal* (10)

In 1989 the Basel Convention on the Control of Transboundary Movements of Hazardous Wastes and their Disposal was adopted and in May 1992 it entered into force. The convention arose out of concern for the international community regarding the transboundary movement of waste in the world. The convention regulates the movement of hazardous wastes across boundaries, and outlines the obligations of its parties with regard to the management and disposal of these wastes in an environmentally sound approach. The convention is intended to offer help in the management of these wastes, monitor traffic of wastes in order to prevent illegal trafficking, encourage cooperation amongst convention parties, as well as develop guidelines for the management of wastes.

2.2 Agreements within the European Union

2.2.1 Convention on the Conservation of European Wildlife and Natural Habitats (11)

The signatories of the Council of Europe in Bern, 1979, signed the Convention on the Conservation of European Wildlife and Natural Habitats. This convention was signed in recognition of the importance of wild flora and fauna, and the importance of cooperating with regard to habitat conservation, especially by governments in terms of national goals and programmes. This convention aims to help signatories to conserve natural flora and fauna and their natural habitats, especially in the case of endangered species. This is to be achieved through national policies, which ensure that the flora and fauna of a particular country are conserved at appropriate levels.

2.2.2 The EC Directives

EU legislation on water is very extensive. The most relevant directives are briefly described below.

Nitrate Directive (91/676/EEC)

The Council Directive on the Protection of Waters against Pollution caused by Nitrates from Agricultural Sources addresses water pollution by nitrates from diffuse agricultural sources (fertilisers and livestock manure). The Member States have to define 'good agricultural practice' to reduce nitrate emissions. Furthermore, they have to identify zones that are especially vulnerable to nitrate pollution.

Bathing Water Directive (76/160/EEC)

The Directive on Bathing Water Quality aims at monitoring and testing bathing water quality in order to protect bathers from health risks and to preserve the environment from pollution. The directive applies to all bathing water, except that found in swimming pools and that used for therapeutic purposes, and includes physical, chemical and microbiological limits, which had to be met by 1985.

Dangerous Substance Discharge Directive (86/280/EEC)

The Dangerous Substances Directive aims at reducing emissions of toxic, persistent and bio-accumulating substances discharged by industries to aquatic systems. The Member States may choose Emission Limit Values or Environmental Quality Standards to reach the targets set by this framework directive. As a result of the Dangerous Substances Directive, emissions limits have been introduced for certain substances, including DDT, carbon tetrachloride and pentachlorophenol.

Urban Waste Water Treatment Directive (91/271/EEC)

The Urban Waste Water Directive includes Emission Limit Values for urban wastewater treatment plants. The Directive aims at reducing water pollution from households and small and medium-sized industries. In a stepwise procedure the Member States are asked to implement appropriate primary, secondary and more advanced wastewater treatment measures by the year 2005. The type of treatment facilities chosen depends on the number of inhabitants and the occurrence of specific 'sensitive areas', which have to be defined by the Member States themselves. Major emphasis is placed on the nutrients nitrogen and phosphorous.

2.2.3 Integrated Coastal Zone Management (ICZM)

To improve the management of coastal zones in Europe, the European Commission has developed a European strategy for ICZM. The strategy builds on and extends existing instruments and programmes and defines the EU's role in providing leadership and guidance to support the implementation of ICZM by the Member States at local, regional and national level. The main instruments are Commission Communication (COM/00/547 – Sept 2000) and ICZM Recommendation (2002/413/EC May 2002). The communication establishes the framework of the activities that the Commission will undertake in promoting ICZM, while the recommendation outlines steps that Member States should take to develop national strategies for ICZM.

National level – The national administration must:

- Provide a legal and statutory framework adequate to enable implementation of ICZM at lower levels of administration
- Ensure coherence of national legislation and programmes that affect the coastal zone — a process that entails the co-operation and involvement of a wide range of sector-based branches of the administration
- Promote a national vision to give guidance and support to promote coherent activities at regional and local level

Suggested strategies:

- Develop or maintain 'land-sea' policies, legislation
- Promote bottom-up initiatives
- Identify sources of long-term funding for ICZM
- Identify mechanisms for full and coordinated implementation of EU policies and legislation
- Include monitoring and information systems
- Determine role training and education in support of ICZM

2.3 Agreements of particular importance to the Baltic Sea Region

The states around the Baltic Sea have signed numerous international and multilateral agreements on environmental protection and nature conservation, some of these are presented in the two previous sections. From all the agreements affecting the Baltic Sea region, two 'actors' and corresponding agreements are selected here for deeper discussion: the Helsinki Commission representing the Helsinki Convention and the European Union. Of the many directives issued by the EU, special attention is given to the Water Framework Directive (WFD) and the Environmental Impact Assessment Directive (EIA).

Some European conventions and their status in Sweden, Germany and Russia can be seen in Table 1.

Tab. 1: European conventions and their status in Sweden, Germany and Russia

Conventions in Europe	**Area**	**Sweden**	**Germany**	**Russia**
The Helsinki Convention	Marine protection of the Baltic Sea	✓	✓	✓
The Copenhagen Agreement	Protection from oil and chemical discharges in the Baltic Sea	✓		
The Gdansk Convention	Fishing and conservation of the living resources in the Baltic Sea and the belts	✓	✓	✓
ACCOBANS	Conservation of terrestrial, marine and avian migratory species in the Baltic Sea area	✓	✓	
The Paris Memorandum	Shipping and ports	✓	✓	✓

2.3.1 The Environmental Impact Assessment Directive (85/337/EEC)

The EU Directive on the Assessment of the Effects of Certain Public and Private Projects on the Environment (85/337/EEC) identifies two groups of development projects, those which are always subject to an EIA and those for which an EIA may be required depending on the likelihood of significant environmental impacts.

The environmental impact assessment requires the concerned parties to identify, describe, and assess, in an appropriate manner, the direct and indirect effects of a project with regard to '*human beings, fauna and flora*; *soil, water, air, climate and the landscape*; *material assets and the cultural heritage*'.

The EIA must be conducted on the basis of appropriate information supplied by the developer. This information may be supplemented by the authorities and by the people who may be concerned by the project in question.

The directive also establishes the specific procedure to be followed when a member state is aware that a project is likely to have significant effects on the environment in another member state.

At the regional level, the County Administrative Board with its planning department is responsible for planning in general. It also coordinates different planning interests between the counties and is the first level of appeal for planning and development issues (13). There are also possibilities to develop regional plans, which can be used when questions concerning water and land use need to be coordinated between several municipalities. The regional plan, although not legally binding, should serve as a foundation for plans at the local level, but the possibility to develop regional plans is seldom utilised.

At the local level, the municipality is responsible for the development of several different plans, namely the comprehensive plan, the detailed plan, area regulations and the property regulation plan. All municipalities must have a comprehensive plan covering the entire area of the municipality. This plan is not legally binding and should be revised once every mandate period. The comprehensive plan should contain the fundamental issues concerning the use of land and water areas, the development and preservation of the built environment, the environmental and risk factors that should be considered and finally how national interests and environmental quality standards are considered.

A detailed plan is legally binding but covers only small parts of the municipality and is not compulsory. However, a detailed plan must be developed when new continuous development is considered and when new individual construction projects such as harbours, which have a significant impact on the surroundings, are planned. A detailed plan is also necessary when considering new construction projects in a wider perspective and when considering developments that will be altered or preserved if comprehensive control is required.

2.3.2 The Water Framework Directive (00/60/EC)

The Water Framework Directive (WFD) is a new overarching directive for water quality. It was adopted in December 2000 and presupposes that all member states have implemented the Nitrate, Drinking Water and Groundwater Directives mentioned above. The Water Framework Directive covers both surface water (inland, transitional and coastal waters) and groundwater. The ecological perspective is emphasised, as the directive refers not only to water quality, but also to the water environment in its entirety. This means that the directive covers all management of the sea (including fisheries).

The main goals can be summarised as follows:

- All waters, including surface waters, groundwater and coastal waters, are to reach a good status by 2015, while water bodies currently of good or high status are to be maintained at that level

- A water management approach based on river basins is to be used. This approach considers not only the water itself but the entire catchment area and thus all factors influencing water quality
- Planned measures to protect water resources are to be selected under the consideration of cost-efficiency criteria.

Unlike before, the water body and its overall ecological quality now constitute the centre of water policy and not only the use of water as a resource. The WFD is important to all environmental pressures on coastal areas, such as protection, water quality and water pollution, including pollution from point sources and diffuse sources.

The ambitious goals of the WFD have to be reached within a relatively short time period (see Appendix I). The transposition into national law was expected to be accomplished by December 2003. By the end of 2004, a status report had to be completed including an analysis of the characteristics of each river basin, a register of protected areas, a review of anthropogenic stresses and an economic analysis of water uses. Based on the status report and the monitoring results, the programme of measures has to be drawn up and published by 2009. The programme of measures has to be operational by 2012 and the environmental objectives achieved by 2015.

2.3.3 The Helsinki Convention

Since 1974, the states around the Baltic Sea have cooperated in the field of environmental protection within the framework of the Helsinki Convention, signed by all states around the Baltic Sea. The executive body is the Helsinki Commission, HELCOM.

The latest Helsinki Convention was drawn up and adopted in 1992 by the nine coastal states framing the Baltic Sea and the European Union as contracting parties. The convention also includes the coastal and internal waters within the 12 nautical mile zone.

Most of the decisions of HELCOM are made as jointly agreed recommendations for implementation by the contracting parties, the EU member states, through their national legislation. No country can be forced to implement these recommendations and consequently there is no mechanism for sanctions if they do not. Therefore, many recommendations are not fully implemented or implemented at all. Nevertheless, the recommendations have had an impact on environmental policies in the EU member states and have encouraged states to follow the example of their neighbouring countries.

The relevant EU member states have started to cooperate to improve the ecological conditions of the Baltic Sea, understanding each other as partners. After

1992, several recommendations have been adopted which relate to nature conservation, and also – or specifically – to the coastal zone.

Since the beginning of the 1980s the Helsinki Commission has been working to improve the Baltic marine environment, largely through some 200 HELCOM Recommendations.

Successes during this period include:

- Lower discharges of organic pollutants and nutrients from point sources
- A 60% overall reduction in emissions of oxygen-consuming substances from the 132 originally identified hot-spots since the early 1990s, with 34 hot-spots deleted from the list
- Fewer beaches closed for bathing, thanks to improvements in the treatment of industrial and municipal wastewater
- Significant reductions in atmospheric nitrogen deposition
- Progress in the phasing-out of leaded petrol (more than 85% of the petrol consumed in the Baltic Sea region is now unleaded)
- Dramatic reductions in emissions of organo-halogen compounds, such as toxic dioxins and furans
- National regulations banning hazardous substances such as PCB and DDT
- Stricter controls on industry (permits are now compulsory for industrial emissions)
- The recovery of seal and white-tailed eagle populations
- Better special legislation to prevent the pollution of the Baltic Sea by shipping, developed together with the International Maritime Organisation (IMO)
- Measures to eliminate all illegal discharges by ships into the Baltic Sea.

2.3.4 Relationship between HELCOM and EU

The influence of the EU on the policy and decision-making level in individual countries is reflected by the fact that in principle, no HELCOM Recommendation is adopted without having been reviewed in a EU perspective to ensure that it is harmonised with corresponding EU Directives. Today, countries are giving much higher priority to fully integrating their environmental legislation with that adopted by the EU and to implementation of the EU environmental legislation than to implementation of non-binding HELCOM Recommendations.

3. National legislation in Sweden, Germany and Russia

As mentioned above, there are many different regulations and policies to protect the Baltic Sea. However, as the conditions and legal systems differ between the countries, the degree of acceptance and implementation of the directives and recommendations for coastal zone management is variable. As a consequence, even though many of the recommendations and policies are implemented in the countries and appear successful in theory, in practice significant improvements in water protection, water quality and water use are lacking.

The Swedish legislation is unique in that it places the responsibility for the environment off the coast on municipalities. However, special knowledge of e.g. marine biology and fisheries management is lacking at the local level. Therefore, it is important to make this type of knowledge accessible to the municipalities. It is also important to remember that County Administrative boards and government agencies have some expertise in coastal issues, for example within fisheries and water quality issues. Expert agencies and County Administrative boards that have access to the most recent research can provide knowledge to coastal municipalities and other local actors. Despite the fact that initiatives from higher authorities may trigger a process of change, the responsibility for, and interest in, addressing the imbalances in the coastal resource management must also be present locally in order for a renewal to take place in practice.

A special problem in German coastal waters is the varying spatial extension of responsibilities. For example, the federal state minister is responsible for the area extending to the coastal sea boundary (12 nautical miles from the coast). The national ministry is responsible in the following Exclusive Economic Zone. There is an urgent demand for a clarification of responsibilities between federal states and national ministries and authorities.

The legislation concerning the coastal zones in Sweden, Germany and Russia is quite different. In Germany there are quite strict regulations, specific for the coastal zones. These regulations concern the three kilometre wide coastal planning zone and the laws concerning the protection of the shoreline and the sand dunes. The protection system in Germany is based on a top-down structure where plans on a lower level must be coherent with plans that are made on a higher level. The system also allows for the state to veto plans made on lower levels. In the Russian Federation, the legal system with effects on coastal areas is under development. These facts could explain many of the differences concerning developments along the coastline of the Baltic Sea.

Similar to Germany, Sweden has rules and regulations concerning the protection of the shoreline but there is, for example, no specific legislation concerning planning in the coastal zone. The same situation applies to Russia. The best way to protect an area along the coast in Sweden is by the creation of a nature reserve, a strategy that provides good environmental protection of the area, but that is rarely used. Creation of a 'Zapovednik' is also a good solution in the Russian Federation when it comes to coastal protection.

Concerning ICZM in the Russian Federation, the biggest problem seems to be the absence of corresponding legislation for coastal zones. Lack of legislation creates a number of barriers, not only for the process of strategic planning and development of coastal zones, but also for the creation of a national ICZM system to be used on a national scale.

According to some experts in the field, a legal definition of coastal zones would allow more precise specification of the property rights, conditions of joint use (the marine part of a coastal area is the federal property), requirements for creation of plans of integrated territorial development and building on the coastal lands and the status of alluvial territories. It would also allow efficient controlling instruments for relations between different users of natural resources to be introduced.

In addition, the country and its citizens need to adapt thinking to a much longer time scale, as well as improving pre-phase planning and communication. Furthermore, the involvement of citizens in the decision-making process is necessary in order to gain public support and promote feelings of responsibility for the decision-taking.

In the Russian Federation the economic objective attracts great attention and the first question always concerns who is going to pay for the measure, solution or protection. If there is no funding and measures cannot be afforded on the municipal level, the authority responsible for e.g. the coastal area cannot complete its missions. However, when comparing the implementation of HELCOM in the Russian Federation and ICZM progress in Sweden and Germany, it is important to remember the lack of structural divisions and expert in the field of ICZM in administrations of Russian coastal municipal territories. There are no positive experiences or any experience at all in organisational and practical dealing with ICZM in individual areas nor in the Russian Federation as a whole.

When decision-making takes place within coastal zones, communication problems occur in different ways in Germany, Sweden and Russia. In general, there is enough legislation in all three countries to provide for sound coastal protection. The problem is more to coordinate and integrate the various activities and responsibilities within the coastal zone.

The supervisory authorities in Sweden have quite a strong influence on the process of issuing permits for environmentally hazardous operations. At the same time, as part of the county administration board, they must defend financial interests in harbour development operations and the associated maintenance and creation of working opportunities in the region. The generally large population in harbour districts in Sweden leads to quite large regional economic dependence on harbour industries, which may create a conflict of interest within the authority. The internal structure of the supervisory authority is thus important, posing the question of how independent the authority's environmental delegation is from its economic and other delegations.

The Swedish system of supervision is characterised by a self-check principle, as well as by transparency and trust between authorities and companies. The self-check model is considered to be efficient by both the industry and the supervisory authorities. However, this model may also lead to an information imbalance in favour of the operator. Considering the limited resources of supervisory authorities, the current system may lead to an incomplete picture of the actual coastal zone, harbour or water situation. The risk of decisions based on incomplete information is especially evident considering the high complexity of natural systems and processes, as well as the long time spans that need to be taken into consideration in the environmental regulation of coastal waters. The competence of the operator, while making the self-check system an attractive option, also makes it vulnerable. It should therefore be ensured that a failure to take environmental responsibility does not give the operator an economic advantage. This may be problematic, as damage may occur in the distant future, making discounted costs of future liability small.

The delineation of institutional interactions presented in this report shows that several different institutions are involved in permits and management decisions regarding harbours. All these institutions have to interact with each other and other institutions on several different matters and decisions in connection with harbour operations.

This nested interaction scheme may lead to an efficient flow of information between all institutions, which is advantageous. On the other hand, in this scheme there is no institution that must particularly and independently defend environmental water quality and quantity interests on a local and/or regional level. The anticipated new catchment authorities could take on such a role, independently defending environmental water quality interests, provided that they have sufficient resources and decision-making authority.

In any case, it is important to relevant legislation enforcement and practical coastal zone water management that some independent authority has the main responsibility for consideration of the long time spans and large spatial scales that may be involved in water regulations and management problems of coastal zones.

In the examination process for permits according to the Environmental Code, the supervisory authority and the Swedish Environmental Protection Agency are parties. Thus, they give their view on whether and under which conditions they think that a permit should be issued. Others concerned may either be other parties or have the power to make statements in court. The Swedish Environmental Protection Agency only engages in cases of principal interest; for new harbour activities this is likely to be the case. The views of the supervisory authority are usually given a rather large weight in the permit decision of the environmental court. If it is clear that the supervisory authority is against permitting the operation, the operator is likely to try to negotiate with the authority beforehand to settle differences in views.

4. Conclusions

Coastal zones play an outstanding role in trade, transport, agriculture, fishing, industrial production and tourism in the Baltic Sea Region. There are several serious threats to sustainable development in these coastal zones. Two of the main threats are conflicting interests and environmental pollution. The environmental threats are complex. Some have historical roots, although the rapid growth of national economies will create new threats. One example of this is oil transport from different terminals around the Baltic Sea. The harmonisation of the different administrative and legal processes in the region is one way to overcome the barriers preventing more positive development.

There are many different regulations and policies to protect the Baltic Sea but there are not the same conditions and legal systems in all countries that have accepted and are attempting to implement the coastal zone management directives. As a consequence, there is a lack of significant improvement in water protection, water quality and water use, even though many recommendations and policies have been implemented in the countries and appear successful in theory. In particular, the differences between the increasingly harmonised EU legislation and the situation in the Russian Federation have to be noted. The greatest problem in the Russian Federation is the absence of corresponding legislation for coastal zones. Lack of legislation creates a number of barriers not only for the process of strategic planning of coastal zone development, but also for creation of an ICZM system to be used on a national scale.

When decisions are made within coastal zones, communication problems arise in different ways in all countries in the Baltic Sea region. It can be concluded that in all EU countries, there is adequate legislation to provide for sound coastal protection. The problem is more to coordinate and integrate the various activities and responsibilities of actors in the coastal zone.

5. References

1 www.euccd.de/infos/EU_ICZM_high_level_forum_2003.pdf (2005 05 29

2 www.lighthousefoundation.org/index.php?id=166&L=1 (2006 01 19)

3 www.baltic.vtt.fi/balticinfo/text_files/chapter7.3.html (2005 12 07)

4 Japan International Cooperation Agency, Ministry of Transport and Communications, Klaipėda State Seaport Authority. (2004). „Study on the Port Development Project in the Republic of Lithuania. Final report, volume I-2-67

5 www.unesco.org/csi/act/russia/legalpro5.htm#1.1 (2005 11 12)

6 „EU focus o coastal zones" http://ec.europa.eu/environment/iczm/pdf/2000brochure_en.pdf (2005 11 30)

7 „A Common Approach to the Implementation of ICZM in the Baltic Region" www.eucc.nl/en/policy/common_approach.doc (2006 01 24)

8 www.un.org/Depts/los/index.htm: (2006 03 16)

9 www.imo.org (2006 03 16)

10 Basel Convention on the Control of Transboundary Movements of Hazardous wastes and their disposal adopted by the conference of the plenipotentiaries on 22 march 1989, www.basel.int/text/con-e.pdf

11 http://europa.eu/scadplus/leg/en/lvb/l28050.htm (2006 09 01)

12 http://europa.eu.int/comm/environment/iczm/home.htm (2006-03 16)

13 Visions and Strategies Around the Baltic Sea VASAB www.cbss.st/partners/vasab/VASAB

6. Appendix I: Definitions relating to Coastal Zones

Coasts are unique natural heritage sites with irreplaceable ecological, cultural and economic resources. These spaces function as protective corridors of land and water and concentrate population, economic activities and resources in a complex environment to be described with a systemic approach.

Coastal areas are commonly defined as geographical areas of transition between land and sea that have not been defined as a zone.

Such areas generally include:

- Inland areas, which affect the oceans mainly via rivers and non-point sources of pollution
- Coastal lands, wetlands
- Coastal waters (estuaries, lagoons and shallow waters)
- Offshore waters, mainly out to the edge of national jurisdiction (200 nautical miles)
- High seas, beyond the limit of national jurisdiction

Coastal spaces can be seen as dynamic, unpredictable and interdependent sets of subsystems in which the land-sea interactions are the origin of very specific environments such as wetlands, estuaries and open sea areas.

The coastal zone does not have a rigid definition and the delimitation of its boundaries in a given area *(zoning)* varies according to political and administrative considerations. The main components of coastal systems, natural processes and human activities interact in these spaces in a complex manner, forming a so-called *ecosocio-system* in which the dynamic nature and the productivity are vital to the economic well-being and development of many countries. In recent decades the combination of increasing human and natural pressures has jeopardised coastal areas in terms of ecological integrity and regions at risk in terms of socio-economic welfare.

7. Appendix II: Definitions relating to Coastal Zone Management

Decision-making for coastal area management means dealing with uncertain conditions related to variations in socio-economic parameters and environmental factors, such as geomorphological changes, ecological processes, climate change, etc. Traditional methods for coastal planning have shown limitations in managing this uncertainty and the complexity of choices connected with an appropriate and rational use of coastal resources. The intrinsic dynamic nature of coastal spaces has been largely ignored by traditional approaches, together with a narrowed vision of problems due to sector-based management practices. The result is decades of uncoordinated management and inappropriate decision-making.

Coastal management as a term came into common use in 1972, with the implementation of the United States Coastal Zone Management Act. The act recognised that a new coastal management approach was needed. Since then, it has been widely recognised that a simple juxtaposition of sector-based approaches to management and land use planning is not appropriate to guarantee the sustainable use of natural resources. Coastal areas require specific management approaches involving a system of relationships among actors who operate directly or indirectly in the coastal zones.

Integrated Coastal Zone Management (ICZM) is the process dealing with current and long-term coastal problems. ICZM is a 'process of governance and consists of the legal and institutional framework necessary to ensure that development and management plans are integrated with environmental goals and are made with the participation of all affected'. ICZM is based on the underlying concept of sustainable development.

ICZM objectives include:

- Maximising economic and social benefits derived from the use of coastal zone resources
- Minimising conflicts (providing conflict resolution mechanisms) and the harmful effects of activities on resources, on the environment and upon each other
- Stimulating and guiding the sustainable development of coastal regions
- Facilitating integrated decision-making through a continuous decision support process.

8. Appendix III: Overview of the EU Water Framework Directive

Year	Legal requirements
2000	Directive adopted
2003	Transposition into national law. Identification of river basin districts and competent authorities. Preparation of draft register of intercalibration sites.
2004	Characterisation of water bodies, including heavily modified water bodies Review of pressures and impacts and identification of sites at risk of not meeting the environmental objective of 'good status' Establishment of register of protected areas Economic analysis of water use Final register of intercalibration sites
2006	Comprehensive monitoring programmes operational
2007	Repeal of some Directives
2008	Publishing of draft river basin management plans which will include a first draft of the classification of water bodies
2009	River basin management plans produced to include final classification of the ecological status of water bodies. Programme of measures for each river basin district
2010	Water pricing policies contribute to environmental objectives
2013	Repeal of some Directives
2015	'Good' status to be achieved. The importance of planning and environmental assessment in implementing the political goals of sustainable development has been underscored by the introduction of the EU directive on *'Environmental Assessment of Certain Plans and Programmes'* (EC/2001/42) and the introduction of requirements for Strategic Environmental Assessment (SEA) into national legislation. Furthermore, planning for sustainable development is now high on the political agenda in both the Nordic countries and in the EU context. In the Nordic context this is, for example, manifested in the action programme for 2001-2004 *'Spatial Planning as an Instrument for Promoting Sustainable Development in the Nordic Countries'*, which complements the Nordic strategy for sustainable development.

Chapter 2

Methods in Environmental Conflict Resolution (ECR)

Ronald Wennersten

1. Background

Planning of coastal areas is carried out in administrative and political processes which are imbedded in a framework of policies, legislation, culture, etc. The way humans and their institutions handle conflicts is changing. Interest has grown dramatically in consensus building, facilitation, mediation and other forms of conflict resolution through assisted negotiation and voluntary settlement. The general idea is to avoid intractable conflicts, which can cause expensive time delays in projects. In the EU countries there is also a focus on participatory processes as a part of strengthening the democratic process. At the end planning effects people and the only way to avoid conflicts is to involve people early in the process.

In this chapter, the term stakeholder is often used. A stakeholder can be any organisation, company, governmental entity, NGO or individual that has a stake in, or may be affected by, a given approach to environmental effects from activities.

There are numerous examples of intractable (i.e. long-running) conflicts, where environmental aspects often play an important role. There are various reasons why conflicts become intractable, but there are some common characteristics:

- Fundamental or deep-rooted moral conflicts
- Fundamental value differences between stakeholders
- Fundamental conflicts about world views, values, principles or societal structures
- Length of time conflict has persisted
- Severe power imbalances between disputing stakeholders
- Potential for coercion or escalation to violence
- Rigidity of positions, high level of hostility
- Strongly held beliefs or positions
- Complexity of issues, interlocking issues
- Frames held by disputants, i.e. how they view the world.

Working through conflicts relating to the environment, natural resources and public land involves using a range of methods for alternative dispute resolution rather than traditional litigation (the conduct of a lawsuit), in which a judge or jury may impose a judgement or make a final determination. Alternative methods of assisted negotiation include facilitation, mediation and conflict assessment, which allow all parties or stakeholders in a dispute to reach a mutually satisfactory agreement on their own terms. The process of mediation, for instance, provides stakeholders with an opportunity to work out their differences and arrive at joint solutions with the help of an impartial mediator. Doing so can save time and avoid many of the costs of traditional legal proceedings. In addition, stakeholders who work toward a shared, positive outcome can often achieve a better outcome than they would have been given in court.

The failure of the legislative and administrative arenas to effectively address environmental conflicts often means that litigation ensues. However there are several drawbacks to the litigation process. First, litigation does not allow for adequate public participation in important environmental decisions. Second, litigation is ineffective for resolving the issues at stake in environmental disputes. Many underlying controversies therefore remain unresolved, and hence more lawsuits emerge in the future.

Advocates of Environmental Conflict Resolution (ECR) suggest that its informal negotiation techniques of ECR have several advantages, including:

- Less risk for the parties than the risk associated with win-all or lose-all litigation
- Reduction in court costs, legal fees, inflationary delays
- An increase in the efficiency of the outcome
- An increased likelihood of achieving a stable agreement for several years.

In short, when all parties are at the table, there is a better chance that all relevant issues will be raised and that the parties will be better placed to make efficient trade-offs and reach decisions that effectively address the substantive nature of the dispute.

Disputes can be classified as upstream, midstream or downstream. Upstream environmental conflicts involve planning or policymaking. For example they may include the creation and implementation of government policy at the national, regional or local level, such as environmental, natural resource, health or safety policy. Midstream environmental conflicts involve administrative permitting, for example, the granting of environmental permits. Downstream environmental conflicts are about compliance and enforcement. They can involve the ways that land is used, the allocation or distribution of natural resources and the siting of industrial or other large facilities. Environmental conflicts can also involve the prevention, cleanup and consequences of pollution.

Conflicts emerge from differences in values and world views, conflicting interests and the uncertainty that surrounds various courses of action. In addition, popular attitudes, political culture, technology, laws, political interest, economics and religion can influence environmental conflicts. It is important to understand that environmental conflicts are not only about facts and lack of information, but that values are also important. Otherwise all environmental conflicts could be resolved by presenting the facts to all stakeholders.

Environmental Conflict Resolution consists of a set of techniques, processes and roles that enable parties in a dispute to reach agreement, or at least to understand the views of other stakeholders, usually with the help of one or more third-party neutrals. Despite the variation in ECR techniques and processes, researchers have identified five characteristics shared by all forms of ECR:

1. Participation is usually voluntary for all participants.
2. The stakeholders and their representatives must be able to participate directly in the process.
3. Any and all participants must have the option to withdraw from the CR process and seek resolution through a more formal process, such as litigation.
4. The third-party neutral must not have independent, formal authority to impose an outcome but rather should help the parties reach their own agreement.
5. The parties must agree to the outcome or the resolution of the dispute. The purpose of the process is to help parties reach their own solutions, which requires their consent to the decision or recommendation.

In many of the ECR processes there is a need for a neutral third party. We believe that universities can play such a role, but it is always important to stress that the universities are not there to solve conflicts, but to create an arena where the stakeholders involved can achieve good results more effectively.

The way ECR is carried out varies depending on the nature of the actual/potential conflict. Below is a general description of ECR processes. In order to implement these processes in practice, different process tools can be used in combination with different fact tools.

2. ECR processes

Environmental Conflict Resolution processes can be organised along a continuum from less formal, interest-based negotiation or consensus-based techniques to more formal, quasi- judicial arrangements. In consensus-based processes, the disputing parties work together to develop a mutually acceptable agreement. Consensus-based techniques require that everyone, not just the majority, agrees with the outcome or decision, unless all parties specifically agree to define consensus as a majority decision.

Moving along the continuum toward quasi-judicial processes, the goal shifts towards making a determination about the issues in dispute and as a result, the nature of the techniques used also changes. Consensus-based techniques are generally used for upstream disputes, whereas quasi-judicial processes are generally used for downstream disputes.

ECR processes are based on the idea of negotiation. Negotiation is simply bargaining – a process of discussion and give-and-take among disputants seeking to find a solution to a common problem. The process can be relatively cooperative, as when both sides seek a solution that is mutually beneficial, or it can be confrontational, in which each side seeks to prevail over the other. Below is a short presentation of some of the more common ECR techniques.

2.1 Consensus-based processes

The most common consensus-based processes used in environmental conflicts are conflict assessment, facilitation, mediation, conciliation, negotiated rulemaking and policy dialogue.

2.1.1 Conflict assessment

Conflict assessment is valuable as a first step in ECR processes. The goal of conflict assessment is to help identify the issues in dispute, the stakeholders affected and the appropriate forms of ECR for handling the conflict. The assessment process typically involves multiple steps, beginning with discussions among potential stakeholders to evaluate causes of the conflict and to identify the entities and individuals who would be substantially affected by the outcome of the conflict. The next step is to assess the stakeholders' interests and needs, identify a preliminary set of relevant issues to be discussed and evaluate the feasibility of using consensus-building or other collaborative process to address the issues in dispute. Other prospective stakeholders are then informed and educated about the possible consensus-building ECR processes so that they can determine whether they want to participate. The final step of the conflict assessment process is to select the ECR technique most appropriate for the conflict. In this step, often called process design, a neutral party recommends or assists in developing a process for addressing a particular controversy or dispute. The process designer typically interviews representatives of the interested or affected groups to understand their perceptions about the conflict, interests in the conflict and their suggestions about useful ways to handle it. With this information, the designer then reports to the parties with recommendations or a plan for handling the dispute.

2.1.2 Facilitation

Facilitation is a collaborative process in which a neutral party assists a group of stakeholders in constructively discussing the issues in dispute. The facilitator typically works with participants before and during these discussions to ensure that the appropriate people are at the table, to help the parties set and enforce ground rules and agendas, to assist parties in effectively communicating, and to help the participants keep on track in working toward their goals. Facilitation may work in any number of situations, ranging from scientific seminars to management meetings to public forums.

2.1.3 Mediation

Mediation is a form of facilitated negotiation in which a skilled, impartial third party with neither decision-making authority nor the power to impose a settlement assists the parties in reaching a voluntary, mutually agreeable resolution to all or some of the disputed issues. Whereas facilitation bears many similarities to mediation, the neutral facilitator usually plays a less active role than a mediator and, unlike a mediator, often does not see resolution as a goal of their work. Typically the process involves supervising the bargaining, helping the disputants find areas of common ground and understand their alternatives, offering possible solutions, and helping parties draft a final settlement agreement.

2.1.4 Conciliation

Conciliation involves efforts by a third party to improve the relationship between two or more disputants. Generally, the third party works with the disputants to correct misunderstandings, reduce fear and distrust and improve communication among the parties in conflict.

2.1.5 Policy dialogue

Policy dialogue is generally used to address complex environmental conflicts or public policy disputes. In the policy dialogue process, representatives of groups with divergent views or interests are assembled to generate discussion and improve communication and mutual understanding. The goal is to explore the issues in dispute to see if general recommendations can be developed and to try to reach agreement on the policy standard or guidelines the government will propose. Unlike the other consensus-based ECR processes, policy dialogue usually does not seek to achieve a full, specific agreement that would bind all participating interests. Rather, participants in a policy dialogue may seek to assess the potential for developing a full consensus resolution later or may put forward general, non-binding recommendations or broad policy preferences for an agency (or other governmental entity) to consider in its subsequent decision-making.

2.2 Quasi-judicial processes

Quasi-judicial processes supply the disputing parties with an expert opinion about the merits of their cases, furnish more information about their best alternative to a negotiated agreement, and provide the disputants with a loop back to negotiation. In theory, the parties can engage more effectively in settlement negotiations if they are provided with additional information on these processes. Some of the most common quasi-judicial processes used in environmental conflicts are *early neutral evaluation*, *arbitration* and *fact finding*.

2.2.1 Early neutral evaluation

Early neutral evaluation is a quasi-judicial process in which a third-party neutral, often someone with specific relevant legal, substantive or technical expertise, hears informal evidence and arguments from all the parties involved in the dispute and issues a non-binding report advising them about the strengths and weaknesses of their cases. The report may also evaluate the likely reaction of a judge or jury if settlement is not reached, provide guidance about an appropriate range of outcomes, and assist the parties in narrowing the areas of disagreement or identifying information that may enhance the chances of settlement.

2.2.2 Arbitration

In arbitration, the disputants present their cases to an impartial party, who then issues an opinion. An arbitration can be binding or non-binding. In binding arbitration, the opinion is final and subject to limited judicial review. In non-binding arbitration, the opinion is advisory and can be rejected; however, it can also serve as a loop back to negotiation.

2.2.3 Fact finding

In this process, a neutral called the fact-finder receives information and listens to arguments presented by the disputants. The fact-finder, who may conduct additional research to investigate the issues in dispute, evaluates the evidence and submits a report that contains findings of facts and sometimes recommendations based on these findings.

3. Process tools

Process tools are very much about creating platforms for constructive discussions among the stakeholders. Below are short descriptions of some process tools that are commonly used in projects and planning activities.

3.1 Scenario Analysis

Scenario analysis is a process of analysing possible future events by considering alternative possible outcomes (scenarios). The analysis is designed to facilitate improved decision-making by allowing more complete consideration of outcomes and their implications.

In term of basic principles, scenarios can be classified as:

- *Forecasting scenarios* – which attempt to encompass future alternative development paths from the standpoint of the current situation and can also include expected or desired policy switches.

- *Backcasting scenarios* – which take as their initial starting point some desired future state of affairs or policy objective and then explore alternative strategies to maximise goal attainment.
- *Descriptive scenarios* – which set out a sequenced set of possible events in a neutral way.

Several attempts have been made recently to use scenarios as a policy-support tool in coastal zone management. The projects most relevant to this report are:

- The Georgia Basin Futures Project (GBFP), a regional integrated assessment project with the purpose of combining public values, preferences and beliefs with expert knowledge in the production of scenarios on the future of the Georgia Basin in western Canada over the next 40 years (Robinson et al. 2001).
- EUROCAT (Project No EVK1-CT-2000-00044), which aims to achieve integrated catchment and coastal zone management by analysing the response of the coastal sea to changes in fluxes of nutrients and contaminants from the catchments. Within the project, the use of scenarios for integrated catchment/coastal zone management in the Humber Estuary, U.K., is investigated (Ledoux et al. 2002).
- Research on integrating structural changes into future research and modelling on the Seine River Basin through studying the consequences of changing the watershed's spatial segmentation by considering a hypothetical future state of the river basin in which the city of Paris suffers from sewage discharge upstream. This is an example of coupling structural changes in socio-economic scenarios with existing bio-geo-chemical models by exploring 'surprise-based' scenarios using a backcasting approach (Kieken 2002).

3.2 SWOT Analysis

A *SWOT Analysis* is a strategic planning tool used to evaluate the **S**trengths, **W**eaknesses, **O**pportunities and **T**hreats involved in a project or in any other situation requiring a decision. The required first step in SWOT analysis is a definition of the desired end state or objective. The objective must be explicit and approved by all participants in the SWOT analysis process. Failure to correctly identify an end state leads to wasted resources and possibly failure of the enterprise.

Once the objective has been identified, SWOTs are identified and listed. SWOTs are:

Strengths: Attributes that are helpful in achieving the objective.

Weaknesses: Attributes that are harmful to achieving the objective.

Opportunities: *External* conditions that are helpful in achieving the objective.

Threats: *External* conditions that are harmful to achieving the objective.

Correct identification of SWOTs is essential because subsequent steps in the process of analysing the objective are all derived from the SWOTs. First, the decision-makers have to determine whether the objective is attainable given the SWOTs. If the objective is NOT attainable a different objective must be selected and the process repeated.

If, on the other hand, the objective seems attainable, the SWOTs are used as inputs to the creative generation of possible strategies, by asking and answering the following four questions:

1. How can we **U**se each Strength?
2. How can we **S**top each Weakness?
3. How can we **E**xploit each Opportunity?
4. How can we **D**efend against each Threat?

Ideally a cross-functional team or a task force that represents a broad range of perspectives should carry out the SWOT analysis.

The following errors have been observed in published accounts of SWOT analysis:

1. Conducting a SWOT analysis *before* defining and agreeing upon an objective (a desired end state). SWOTs should not exist in the abstract. They can exist only with reference to an objective. If the desired end state is not openly defined and agreed upon, the participants may have different end states in mind and the results will be ineffective.
2. Another error is to confuse SWOTs with possible strategies. SWOTs are descriptions of conditions, while possible strategies define actions. This error is made especially with reference to opportunity analysis. To avoid this error, it may be useful to think of opportunities as 'auspicious conditions'.

3.3 Logical Framework Analysis

Logical Framework Analysis (LFA) is a tool for planning and managing development projects. It takes the form of a table (or framework) and aims to present information about the key components of a project in a clear, concise, logical and systematic way. A logical framework summarises, in a standard format:

- What is the project going to achieve?
- What activities will be carried out to achieve its outputs and purpose?
- What resources (inputs) are required?
- What are the potential problems that could affect the success of the project?
- How will the progress and ultimate success of the project be measured and verified?

LFA can be a useful tool in the planning, monitoring and evaluation of development projects. It is not the only planning tool, and should not be considered an end in itself, but using it encourages the discipline of clear and specific thinking about what the project aims to do and how, and highlights those aspects upon which success depends. LFA also provides a handy summary to inform project staff, donors, beneficiaries and other stakeholders, which can be referred to throughout the lifecycle of the project. LFA should not be set in concrete. As the project circumstances change, it will probably need to reflect these changes but everyone involved will have to be kept informed. There is a general perception that using LFA is intimidating, perhaps because of awareness of the complexity of development projects. However, the essentials of any conflict can be reduced to one or two sides of A4. It should be remembered that the logical framework is not intended to show every detail of the project or to limit the scope of the project. It is simply a convenient, logical summary of the key factors of the project.

3.4 Hearings

Hearings create a platform where an issue can be presented and different stakeholders can ask questions around anything relating to this issue. A hearing usually takes the form of a meeting or a series of meetings. Sometimes the hearing process starts with a pre-hearing meeting to reduce the number of outstanding issues before the hearing and consequently the time needed for the hearing itself. Hearings can be held with different stakeholder groups and the groups should be selected carefully to ensure that people can speak openly.

3.5 Role play

Role play is a rarely used qualitative teaching technique mostly used in courses in universities. Role play encourages participants to reflect on a potential conflict and is an important process in trying to understand other stakeholders' opinions and frameworks. In role play, participants adopt and act out the role of stakeholders or interests that may have personalities, motivations and backgrounds different from their own.

A role play consists of three major components. First, there is a briefing session, in which participants are informed about the nature and purpose of the activity and, if needed, the players who are to act out the roles are briefed about the situation and setting for the role play. Non-players can be given specific observational and reflective learning tasks. Second, the role play is enacted. Third, and of enormous importance for the learning process, a debriefing session is undertaken in which the experience is discussed and explored.

The role play developed by the original author can be very general or a more specific scripted piece of role play. It should be noted that the script is not a sequence of dialogue to be reproduced, as in a theatre play, but rather a sequence of instructions of the order of interaction between players, with the occasional line of dialogue to ensure that key statements are made. A role play was used in the pilot course which is described I another chapter.

3.6 Development of Sustainability Indicators

An indicator is something that helps to understand the current position and determines the direction of change and the distance from the desired end point. A good indicator alerts the parties involved to a problem before becomes too serious and helps them identify what needs to be done to solve the problem. For example, indicators of a sustainable community point to areas where the links between the economy, environment and society are weak and help show the way to rectify the situation.

Sustainability indicators are different from traditional indicators of economic, social and environmental progress. Traditional indicators – such as stockholder profits, incidence of asthma or water quality – measure changes in one part of a community as if it were entirely independent of the other parts. Sustainability indicators reflect the reality that the three different segments are very tightly interconnected.

Indicators are as varied as the types of systems they monitor. However, there are certain characteristics that effective indicators have in common:

- Effective indicators are *relevant*; they reveal essential information about the system
- Effective indicators are *easy to understand*, even by non-experts
- Effective indicators are *reliable*; the information that the indicator provides is trustworthy
- Effective indicators are based on *accessible data*; the information is available or can be gathered while there is still time to act.

There are no perfect sustainability indicators, but there are indicators that address the critical issues of community sustainability. These indicators help analysts understand and measure progress better than traditional indicators. In this section, indicators that are currently in use by a community group or organisation actively working on sustainable community issues are highlighted. The purpose of highlighting a particular indicator is not to suggest that every community should be using it, but rather to provide food for thought and promote discussion of the issues involved in developing indicators that measure and encourage progress towards becoming sustainable communities.

Sustainability indicators should:

1. Address the issue of the community's carrying capacity relative to the four types of community capital: natural, human, social and built.
2. Highlight the links between the community's economic, social and environmental well-being.

3. Focus on a long range view.
4. Be understandable to the community.
5. Measure local sustainability that is not at the expense of global sustainability.

3.6.1 Short overview of indicators for integrated coastal zone management (ICZM)

On a global level, coasts comprise 20% of the Earth's surface, yet they host a significant proportion of the entire human population (approximately 50% of the human population lives within 200 km of the coast (UN 2002)). Coastal ecosystems are highly productive, containing high biological diversity, rich fishery resources and significant seabed minerals. Coasts also support a diverse array of related industries (e.g. fisheries and aquaculture, tourism, shipping, oil and gas industries), which provide enormous economic productivity.

However, the increasing demands of human activities and living places in coastal zones are placing stresses on finite coastal systems and resources. For example, at a global level, 48% of fish stocks are fully exploited and 28% are depleted, overexploited or recovering (FAO 2001). Water quality is affected by pollution from ships (GESAMP 2001) and pollution from land-based sources (e.g. intensification of agricultural practices is contributing to the impact of nutrient loading and eutrophication of estuaries and bays). Use of fossil fuel continues to cause global climate change, which in turn is causing changes in coastal ecosystems.

These and other similar issues are causing widespread concern about the status of coastal zones. Changes in the condition of natural systems have a direct impact on the ecosystem functions that humans depend on for health, services and economic growth. One of the ways to better understand the complexity of such linkages is an indicator-based approach to management.

Indicators can be used to enhance the understanding of the impacts of integrated coastal policies by providing a simplified view of a more complex phenomenon. They provide insights into trends or events that cannot readily be observed (Cummins et al. 2004). Thus, indicators both quantify and simplify information. Indicators characterise a stable environmental state in the real situation and reflect changes in long-term and short-term perspectives, and they also characterise measures for bringing the environment back to sustainable development (Belousova 2000). Indicators could be designed to determine programme performance, or created to establish links between anthropogenic activities and ecosystemic health or to track trends and conditions in ecosystem dynamics or resource use (Bowen & Riley 2003).

Indicators used to monitor and evaluate the performance of an ICZM initiative can be defined as environmental, socio-economic and governance indicators

(UNESCO 2003). Thus far, much of the application of indicators has focused on environmental conditions, typically describing the physical or biological state of the coastal environment. Socio-economic indicators have been developed but are more common at the sub-national (e.g. Atlantic Living Coastlines EU Demonstration Project) than the national level (UNESCO 2003). The use of governance indicators to examine the performance of ICZM is still very much in the early stages of development.

In ICZM, indicators can be used in two basic, overlapping ways; as a means of communication and as a means of measurement. These uses includes informing decision-making, increasing understanding of important issues, assessing conditions and trends, measuring performance and results of policies or actions, and showing the connection between environmental, social and economic concerns (Scottish Executive Central Research Unit 2001).

ICZM requires clear indicators of sustainability that measure the 'health' of the coast in relation to environmental, social and economic activity, whilst using fewer resources (materials, fuels or land), and whilst managing the impacts of the resources that are used so that their adverse effects are minimised (ENTEC 2001). Such indicators are essential tools for monitoring the state of the coastal environment and informing managers and policymakers of the effectiveness of strategies in achieving sustainability. These indicators need to be based on rigorous scientific, social and economic research (Gallagher 1999).

The types of indicators used for monitoring the status of a coastline often depend on the scale of the area under consideration and the target audience. Indicators for small regions tend to address issues in more detail and are often aimed at helping local communities resolve specific local issues. National indicator sets are broader, more general and designed for reporting the progress in achieving targets of different coastal management programmes or national policies.

On 30 May 2002, the European Parliament and the EU Council adopted Recommendation 2002/413/EC concerning the implementation of Integrated Coastal Zone Management (ICZM) with the aim of fostering the development of integrated management strategies to guide European coastal zones toward more sustainable scenarios (EU 2002). To implement this recommendation, the EC created an expert group on ICZM that drew up a list of twenty-seven indicators, which cover natural coastal processes as well as social environment and economic issues. This set of indicators can be used as a basis to help European countries evaluate and improve their coastal zone development and make it more sustainable.

Indicator sets should be adapted for use in each region separately, taking into account differences in national economics, culture and natural environment. Such indicator-based assessment of the state of the coastline should improve the development of the coastal environment, increase the quality of decision-making and ensure the welfare of coastal communities.

Indicators should reflect both the actual situation concerning e.g. environmental status but also opinions about relative importance of different factors. Indicators have therefore to be developed both in top down and bottom up processes where stakeholders are involved in selection of indicators.

4. Fact tools

Fact tools are used to collect facts for use in the ECR process. It is important that these facts are collected in an objective way and presented in such a way that all stakeholders can assimilate and feel trust in the results.

4.1 Inquiries/Interviews

There are two basic research methods: quantitative and qualitative. Each method approaches a research topic differently.

Quantitative research was originally developed in the natural sciences to study natural phenomena. It involves systematic investigation of quantitative properties and phenomena and their relationships. It is used in both the natural and social sciences, including physics, biology, psychology, sociology, geology, education and journalism.

Qualitative research was developed in the social sciences to enable researchers to study social and cultural phenomena. It has special value for investigating complex and sensitive issues. It involves the use of interviews, documents and participant observation to understand and explain social phenomena.

An *inquiry* is any proceeding seeking truth, information or knowledge. Two types of inquiries are presented in the following: case studies and surveys.

A *case study* is an empirical inquiry that investigates a phenomenon within its real-life context. It can include qualitative as well as quantitative data and relies on multiple sources of data. It benefits from the prior development of theoretical assumptions. Case study is not synonymous with qualitative research. It can be based on any mix of quantitative and qualitative data.

Surveys are mostly used to collect quantitative information about items in a population. Surveys of human populations and institutions are common in political polling and health, social science and marketing research. Surveys normally focus on opinions and factual information.

Different techniques are used for data generation and data collection in case studies and surveys. Two such techniques are described here:

A *questionnaire* (also known as self-administered survey) is a type of statistical survey handed out to a specific group of persons to gather information.

Interviews are systematised conversations between individuals and they can be of two types:

Exploratory interviews (also known as in-depth interviews) have the main purpose of developing ideas and research hypotheses rather than of gathering facts and statistics. They are concerned with trying to understand how ordinary people think and feel about the topics of concern. They are not appropriate for data collection, but for collection of ideas.

Standardised interviews have the main purpose of data collection. They are used, for example, in public opinion polls and market research.

Both types of interviews require interpersonal skills, for example putting the respondent at ease, asking questions in an interested manner, noting down the responses, without upsetting the conversational flow, and giving support without introducing bias.

There are mainly two types of questions used in questionnaires and interviews: *Closed questions* and *open* or *free-response questions*. Closed questions are those in which the respondents are offered a choice of alternative replies, while *open* or *free-response questions* are not followed by any kind of predetermined choice, and the answers have to be recorded in full.

Questionnaires have advantages over some types of surveys in that they are cheap, do not require as much effort from the questioner as verbal or telephone surveys and often have standardised answers that make it simple to compile the data.

On the other hand, interviews are better when there are numerous open-ended questions. Such open-ended questions are important in allowing the respondents to say what they think with greater richness and spontaneity. Interviews may improve the response rate and are also useful for reaching less well-educated respondents and those with reading or language difficulties.

4.2 Frame Analysis – A practical approach

One of the problems with multiple uses of natural resources is the creation of environmental conflicts. It is not unusual for high-risk industries to share the same environment as other economic activities and local residents. Such industries pose risks to their neighbours and to the environment in the form of pollution and accidents, and often lead to conflicts. Complex information, uncertainty about consequences and intricate negotiation dynamics contribute to the creation and aggravation of such conflicts. How to avoid them is a question that has moved scholars and decision-makers.

Several practices have been developed and applied to solve problems created during decision-making processes. Although beneficial, these practices have not been entirely successful in solving conflicts, especially when related to sustainable development, environmental degradation, resource depletion and conservation of energy (Schön & Rein 1994). Numerous analyses have shown that such conflicts tend to be persistent. Analysts have suggested that the persistency of the conflicts results from the way the parties select the facts for their attention and/or the way they interpret them. Kheel (2002), a conflict mediator, explained that conflicts are often so difficult to solve because the parties involved do not fully understand what the issues are about; they see things only from their own point of view. Not seeing the other party's position in the situation is what often leads to the intractability of many conflicts. To deal with such difficulties, frame analysis was developed and introduced into conflict resolution.

4.2.1 The concept of frame analysis

Frame analysis was developed based on the concept of framing. It means that the underlying structure of belief, perception and appreciation held by a person will define the way she or he sees the situation, the way the person *frames* a situation (Schön & Rein 1994).

Frame as a concept has become popular among social scientists with the recognition that two or more people involved in the same situation see it or define it in different ways. The way a person frames a situation defines which aspects of the situation the person chooses to observe, i.e. how the person sees the issue determines the way she or he selects or ignores matters for attention (Pinkley & Northcraft 1994).

Individuals frame a situation to help them make sense of the world around them, to *make sense* of the complex reality, abounded by complex and often divergent information about the social world. An individual constructs frames when trying to understand new complicated, sometimes complex, events; to identify and interpret specific aspects that seem to them key to understanding the situation, and to communicate that interpretation to others (Schön & Rein 1994; Lewicki et al. 2003). Through this process individuals select the issues they give attention to, 'imparting meaning and significance to elements within the frame and setting them apart from what is outside the frame' (Buechler 2000, p. 41).

4.2.2 Framing conflicting situations

The processing of framing may or may not construct a conflict situation. A conflict situation is defined as parties observing different aspects of the same situation, or observing the same aspects but understanding them differently. This means that the parties frame the situation in a conflicting way, they hold *conflicting frames* (Schön & Rein 1994).

Understanding how conflicts are created, how they progress and how to effectively manage them are not easy tasks. There are many factors contributing to the difficulties. It is not uncommon that the people in a conflict do not know exactly what the dispute is about or that they experience the same conflict in different ways. It does not mean, however, that 'objective reality does not exist, only that disputants' subjective experience is their reality and thus determines the nature of the conflict to them' (Pinkley & Northcraft 1994, p. 193).

The many difficulties present during a negotiating process emphasise the importance of parties in a dispute being aware of the potential effects of frames, for the simple reason that frames filter reality and limit attention to situational details, and often exclude the possibilities of each party seeing the other's position in the dispute. In this sense, knowing what the frames are and how they are constructed permits understanding of how they affect the progress of the conflict and how they can be used to influence it (Schön & Rein 1994).

4.2.3 Frame analysis

Frame analysis is possible since the frame of parties in conflicts can be deduced from understanding the context in which the conflict is embedded; by asking the parties about their frames; by listening to or reading their communications; and by watching their behaviour. In addition, each individual can 'also try to understand our own frames, by thinking about what aspects of a situation we are paying attention to, emphasising, focusing on, or ignoring – and by observing our own words and actions' (Lewicki et al. 2001, p. 52).

Frame analysis of environmental conflicts has indicated a predominance of 'identity frames' and 'characterisation frames' in environmental disputes. Identity frames focus on how individuals answer the question Who am I? This type of frame is fundamental to the understanding of environmental disputes, since conflicts almost inevitably arise when people's identities are being threatened. On the other hand, characterisation frames mirror identity frames in that they are statements made by individuals about how they understand someone *else* to be; that is, Who are they? (Gray 2003, p. 23). One example of the role of political identity frames and characterisation frames in shaping conflict is presented in the Appendix.

Table 1 below shows the frames mostly found in environmental conflict resolution and the main factors influencing their formation, according to Gray (2003, pp. 11-34).

Various other types of frame underlie the origins of conflicts. For instance, divergent frames of risk perceptions not only directly affect preferences for specific decisions but also aggravate disparities between how people understand their identities in a particular conflict and how they characterise others in the conflict (Elliot 2003). The importance of risk perception frames influencing the progress of a conflict is also presented in the Appendix.

Tab. 1: Frames and factors influencing their formation

Frame categories	Definition	Types of identity	Factor influencing the frames
Identity	Focuses on how individuals understand how they are, the individual's self perception	Social group Institutional Place-based Interested-based	Values and beliefs Social and cultural experience Interests Relational power Organisation of the parties Membership, role, association Demographic characteristics
Characterisation	Statements made by individuals about how they understand someone else to be	Positive frames Negative frames	Experience Belief, prejudices and judgement Individual's social identity: as group identity and group membership Understanding other stake-holders' actions, interests and positions Relationships among stakeholders
Conflict management	Deals with disputants' preferences for how the conflict should be managed or dealt with. Frames selected will define the strategy used by the parties to deal with the conflict	Avoidance/passivity Joint problem-solving Authority decides based on expertise Adjudication Appeal to political action Appeal to market economy	Experience Knowledge of the problem Local experience

Another type of frame of great interest is the action frame. Action frames inform policy practice, frames that are implicit in the agreements that determine the content of laws, regulations and procedures. In this section of the report, action frames, identity frames and characterisation frames, as well as risk perception frames and their roles in the formulation of public policy, are illustrated in the Appendix.

The practice of frame analysis for conflict requires: a) placing the conflict in a historical and political context; b) understanding the physical and organisational settings that led to the conflicts; c) identifying interested parties, the stakeholders, directly or indirectly involved in the conflict; and d) analysing the parties' positions in relation to the conflicts.

4.2.4 Frame reflection

Environmental conflict analyses have generally been conducted in retrospect or in parallel with the ongoing decision-making process. This also holds for retrospective frame analyses, which have been conducted in retrospect or in parallel. These analyses provide insights into the process of frame construction, leading to understanding of how the different parties construct their frames and how these frames influence the progress of the conflicts. The importance of such analyses is to a great extent educational; they provide insights to the practitioners and researchers and eventually strengthen their capacity for conflict resolution. However, retrospective analyses do not offer participants in a dispute an opportunity for reflection about the process in which they are currently engaged and which could benefit from the frame information given, which is what a frame reflection does.

Frame reflection is based on the notion that once parties in a dispute can get a better understanding of their frames and that of others, they can begin to shift their often tacit and untested ways of seeing the world and, consequently, the issues in conflict. This can lead to a reframing of the conflicting issues. More generally, existing frames can be altered by efforts to improve communication and interaction in a reflective way, especially if the parties are conscious of the process of framing. However, the way to go from frame analysis to conflict resolution is not an easy task; it is in fact complicated and complex. It requires specific training and experience, enabling frame analyses/frame/reframing conflicting issues to first be conducted and finally achieving the aim of conflict resolution.

Some recommendations may help smooth the way. These are (Schön & Rein 1994, p. 195):

1. Learn from best practice: Researchers and policy practitioners might undertake projects aimed at leaning from success, carefully studying examples of policy practice identified as most effective.
2. Learn from retrospective study of intractable conflicts: Academics and practitioners might collaborate in the retrospective study of intractable conflicts, tracing back the communicative interactions central to the conflicting issues.
3. Conduct frame-reflective studies of policy-making-process: Academics and practitioners might collaborate in conducting frame-reflective studies of policy-making in process, seeking to identify points of blockage that lead the actors to a stalemate.

Academics can have two main functions. First they can help the practitioners with the difficult process of frame reflection; they can help the practitioners construct the frames that underlie their own positions in the conflict and that of the others, for example, by helping them to construct how the others see and interpret the conflicting issues. From this starting point, academics might go on to help practitioners reframe a conflicting issue. Second, academics can help practitioners to create conditions for mutual trust, including repairing breaks of trust that so commonly happen under conditions of threat and pressure.

A case study of a frame analysis of an oil spill that occurred in 2000 in Guanabara Bay, Brazil, is given in the Appendix. It was developed with the main purpose of investigating the usefulness of frame analysis in the study of environmental conflicts. Tools (a)-(d) listed above were used to carry out the frame analysis. The application of these tools required much time and effort, but was essential for the understanding of the conflicts. This case study is an example of a downstream ECR, but frame analysis can also be used in upstream ECR where the different stakeholder frames are analysed in order to avoid conflicts.

4.3 Risk Assessment

Risk Assessment is about trying to understand and evaluate different kind of risks associated with an activity. Risk is by nature uncertain and the uncertainty can be related e.g. to how probable an accident is and/or what the consequences would be if the accident occurred.

Consequences can affect the environment, humans and property. Risks are connected with accidents or with continuous exposure of environment and humans. For example, when handling oil in terminals and ships there is always a small amount of spillage. These types of risks are usually handled in the Environmental Impact Analysis.

Where there are large risks for accidents involved, e.g. storage of oil and gas or chemicals in large amounts, EU legislation (SEVESO Directive) states that there has to be a document called a safety report produced by the company responsible for the activity. This document should also contain a Worst Case scenario. The problem is how to describe this scenario. There is no actual support for this in the SEVESO guidelines.

The main problems associated with risk assessment relating to large-scale accidents are:

How probable is the accident?

Estimating the probability of an accident often involves complicated calculations with a high degree of uncertainty. The regulations do not state in detail which types of accident should be taken into account. Usually the company chooses not the Worst Case, but a less catastrophic but more likely event. The calculations involve estimations of both technical and human failures.

What will be the consequences?

The consequences will depend very much upon which scenario is chosen, a Worst Case or a less catastrophic but more likely event as mentioned above. The most difficult task is usually to estimate the short-term and long-term consequences for the environment.

Can the situation be handled in an appropriate way?

Companies should choose the actual Worst Case, not the more likely but less catastrophic worst plausible case. It is better for companies to have the external community complain that they are worrying too much about vanishingly unlikely scenarios than to be blamed for ignoring such scenarios.

In general the conclusions in the safety reports are not readily accessible to non-experts. This does not mean that they are complicated and correct. There is actually very much experience showing that the results are generally highly uncertain.

Some general recommendations from risk assessment in participatory processes are:

1. Do not understate the Worst Case, in content or in tone. If companies stress the low probability of this scenario, the external community will stress its high magnitude. Better to stress its high magnitude, and let the community stress its low probability.
2. Acknowledge that risk assessment of catastrophic risks is extremely uncertain, and in key ways less conservative than risk assessment of chronic risks. Accident scenarios that were missing in the calculations quite often occur.
3. Pay more attention to improving mitigation and prevention than to estimating magnitude and probability. The 'right' debate is over what should be done about disaster risks, not whether they are quantified correctly.
4. If some stakeholders have played an important role in getting a company to open up its disaster planning, the company should give them the credit they are due and involve them in planning and communication.
5. Some of the regulations for communicating about Worst Case scenarios may seem foolish. Companies should not use foolish regulations as an excuse for making foolish communications. They should do what is required and what is sensible. They should also listen carefully to stakeholders' opinions, as there is usually some truth and some additional information contained in them.

4.4 Environmental Impact Analysis (EIA)

Environmental Impact Assessment (EIA) is used to identify the environmental and social impacts of a project prior to decision-making. It aims to predict environmental impacts at an early stage in project planning and design, find ways and means to reduce adverse impacts, shape projects to suit the local environment and present the predictions and options to decision-makers. By using EIA, both environmental and economic benefits can be achieved, such as reduced cost and time of project implementation and design, avoided treatment/clean-up costs and impacts of laws and regulations.

Key Elements of EIA include:

- Scoping: Identifying key issues and concerns of interested parties
- Screening: Deciding whether an EIA is required based on information collected
- Identifying and evaluating project alternatives: Listing alternative sites and techniques and the impacts of each
- Mitigating measures dealing with uncertainty: Reviewing the proposed actions to prevent or minimise the potential adverse effects of the project
- Issuing environmental statements: Reporting the findings of the EIA.

4.5 Strategic Environmental Assessment (SEA)

Strategic Environmental Assessment (SEA) can be seen as a complement to Environmental Impact Assessment (EIA), which is mainly performed on a project level. It is usable earlier in the decision-making process. SEA can be seen as the environmental assessment of a strategic action, and strategic actions are defined as policies, plans and programmes. Different countries define this method in different ways.

Strategic Environmental Assessment demands strategic actions that are considered to result in large environmental impact. SEA on policies, plans and programmes are performed at varying degrees of detail. Policies are often much more vague in their presentation and may cover broad areas (geographical and social), whereas some programme SEAs may be performed at almost the same degree of detail as project EIAs.

While analysing the EIA process, some issues for which EIA cannot be used have been identified, for example the effects of broad economic activities in the context of a whole region. In order to solve this problem, the Directive of the European Parliament and of the Council 2001/42/EC of 27 June 2001 on the Assessment of the Effects of Certain Plans and Programmes on the Environment was adopted, which serves as basis for strategic EIA.

The Directive sets the minimum requirements for carrying out strategic assessment of effects on the environment, as well as plans and programmes defined, for which this is a compulsory demand. These fields/sectors are:

- Agriculture
- Forestry
- Fisheries
- Power industry
- Industry, transport
- Waste management
- Management of water resources
- Telecommunications

- Tourism
- Territory planning measures, which are mentioned in the Annexes of the EIA Directive 97/11/EC.

5. Conclusions

The general experience is that areas of potential conflict should be handled as early as possible and with great openness. If information is concealed in order to avoid discussions, the road to results might turn out to be much longer, not shorter. There is usually a general eagerness to get the process underway, but the time needed for information and discussions must be set aside at an early stage. This tends to let stakeholders feel that their opinions have been taken into account, even if the final solution is somewhat different from what they would have wished.

One negative example of conflict resolution strategies is the planned gas pipeline at the bottom of the Baltic Sea, which Russia is planning to build to Germany. The information about the project came out late and many stakeholders feel the information process is not very open. If the information around the project had started much earlier, the opposition to the project would probably have been much weaker and the project group could have taken care of important stakeholder opinions.

Exploration of energy sources might turn out to be one of the most difficult areas for conflicts in the Baltic Sea Region. In order to solve these conflicts, there will be a need for platforms for information exchange and open discussions at early stages. One such platform could be joint Baltic Sea Region courses in conflict resolution strategies, such as that developed in the COASTMAN project. Another experience is the importance of a third party in the conflict resolution process. This third party could be a university who should play a neutral role and establish platforms for conflict resolutions.

6. References

Belousova, A.P. (2000). A concept of formatting a structure of ecological indicators and indexes for regional sustainable development. Environmental Geology 39 (11), 1227-1236.

Bowen, R.E. & Riley, C. (2003). Socio-economic indicators and integrated coastal management. Ocean and Coastal Management 46 (3), 299-312.

Buechler, S.M. (2000). Social movements in advanced capitalism. New York: Oxford University Press.

Cummins, V.O., Mahoney, C. & Connolly, N. (2004). Review of the Integrated Coastal Zone Management Principles of Best Practice. Prepared for the Heritage Council by the Coastal and Marine Resources Centre. 86 pp.

Elliot, M. (2003). Risk perception frames in environmental decision making. Environmental Practice 5, 214-222.

ENTEC (2001). Sustainability Indicators for Waste, Energy and Travel for Scotland. Scottish Executive CRU, Edinburgh.

EU (2002). Recommendation 2002/413/EC of the European Parliament and of the Council of 30 May 2002 concerning the implementation of Integrated Coastal Zone Management in Europe.

FAO (2001). The State of World Fisheries and Aquaculture 2000. Rome: FAO.

Gallagher, A. (1999). Indicators for Sustainable Development on the Coast. Their Role in the Integrated Coastal Zone Management of Devon and Cornwall. Southampton Institute. Southampton.

GESAMP (2001). Protecting the Ocean from Land-Based Activities. Reports and Studies 66. London: GESAMP.

Gray, B. (2003). Framing of environmental disputes. In: R. Lewicki, B. Gray and M. Elliot (eds.) Making Sense of Intractable Environmental Conflicts: Concepts and Cases (pp. 11-34). Washington DC: Islands Press.

Kheel, T.W. (2002). Understanding the art and technique of conflict resolution. Negotiation Journal (October), 304-306.

Kieken, H. (2002). Integrating structural changes in future research and modelling on the Seine River Basin. Research Group on Management of Environment and Rural Areas (RGTE), International research Centre on Environment and Development (CIRED), French Institute of Forestry, Water and Environmental Engineering (ENGREF), Paris, France

Ledoux, L., Beaumont, N., Cave, R. & Turner, K. (2002). Towards integrated catchment/coastal zone management: Science, policy and scenarios. CSERGE Working Paper ECM 02-05.

Lewicki, R.J, Gray, B. & Elliot, M. (eds.) (2003). Making Sense of Intractable Environmental Conflicts, Concepts and Cases. Island Press, Washington, Covelo and London.

Oppenheim, A.N. (2003). Questionnaire Design, Interviewing and Attitude Measurement. Continuum, London, New York. First published as Questionnaire Design and Attitude Measurement by Basic Books in 1966.

Pinkley, R.L. & Northcraft, G.B. (1994). Conflict frames of reference: Implications for dispute processes and outcomes. Academy of Management Journal 31 (1), 193-205.

Robinson, J., Rothman, D., Tansey, J., VanWynesberghe, R. & Carmichael, J. (2001). The Georgia Basin Futures Project: Bringing together expert knowledge, public values, and the simulation of sustainable futures. Georgia Basin Futures Project. Working Paper. October 18, 2001.

Schön, D.A. & Rein, M. (1994). Frame Reflection: Toward the Resolution of Intractable Policy Controversies. New York: Basic Books.

Scottish Executive Central Research Unit (2001). Indicators to Monitor the Progress of Integrated Coastal Zone Management: A Review of Worldwide Practice. 50 pp.

UN (2002). UN Atlas of the Oceans. www.oceansatlas.org/

UNESCO (2003). A Reference Guide on the Use of Indicators for Integrated Coastal Management – ICAM Dossier No.1. IOC Manuals and Guides No.45, UNESCO/IOC.

7. Appendix: A case study on framing environmental conflicts

Industrial development has greatly contributed to economic prosperity and improved standard of living, but it has also brought new risks to people's everyday lives and to the environment. This fact has been evident in the use of environmentally sensitive ecosystems. It is not unusual for high-risk industries to share the same environment as other economic activities and local residents. Such industries often pose risks to their neighbours in the form of pollution and accidents.

Guanabara Bay is an example of multiple uses of an environmentally sensitive area. The bay is situated on the southern Brazilian coast, in the state of Rio de Janeiro. It is an important economic centre and due to its beauty it has become an international tourist destination. For many, Guanabara Bay is the symbol of Rio de Janeiro. It is where the city was founded and grew. The bay is an example of natural beauty combined with a blatant abuse of the fragile natural ecosystems by industrial enterprises. Today the bay is one of the most densely populated areas in Brazil. Industries, oil refineries, oil terminals and shipyards share the same environment as fishing, tourism, commerce and many other economic activities. The oil spill that occurred in Guanabara Bay in 2000 was selected for this case study due the characteristics of the technological risks and the peculiarities of the surroundings where the accident occurred.

History of the oil industry in Brazil

The Brazilian oil company Petrobras is Brazil's largest company and one of the 14 biggest oil industries in the world. The company was created in 1953 as a result of the federal government's decision to develop the oil sector in the country. It was a nationalist project that gave great power to the state for the establishment and operation of the oil sector. A decade later, a military dictatorship took office and administered the country for more than 20 years. The oil sector was among the priorities of the military government, and therefore financial incentives and autonomy were given to the newly created company.

In the late 1970s, the average Brazilian production was 200 000 of barrels of oil per day (bpd), while consumption reached 1 115 000 bpd. In 1990s, Petrobras became the world's main deepwater producer. Today, 75% of the Brazilian oil resources are located in deepwater areas. Over more than four decades, the state-owned Petrobras has become the country's leader in the distribution of oil products, and has steadily increased production. In 2006 Brazil achieved self-sufficiency in oil with national production equal to national consumption. In 2007, the company employs 46 000 workers and 200 000 sub-contracted workers. During these years, Petrobras has reached a technological level comparable to many other international oil companies.

The rapid expansion of the oil sector has been accompanied by severe environmental damage. Pollution and frequent oil spills from oil exploitation and shipping in coastal zones, as well as offshore activities, have had a significant impact on the coastal ecosystems of Brazil. The situation has been especially serious in the densely populated south-eastern part of the country where the largest part of the country's oil production is located. At present, Rio de Janeiro state contains about 80% of the country's total production.

The Refinery Duque de Caxias, REDUC, is the most complete and complex of Petrobras' refineries and the second biggest in terms of capacity. It is situated about 30 kilometres from the city of Rio de Janeiro. The refinery started operating in 1961 with only six plants and a processing capacity of 89 000 bpd. REDUC derived great benefits from the incentives given during the military government to refining, commerce and transport of imported oil. Starting from six plants, today the refinery has 30 plants for oil processing and the capacity to process 242 000 bpd. The refinery occupies an area of 13 km^2, corresponding to 2 000 football pitches.

The oil spill

On 18 January 2000, 1.3 million litres of oil leaked out into the adjacent waters of Guanabara Bay. The leak was due to a rupture in a 16.4 km long pipeline transporting marine fuel from REDUC to the oil terminal at Ilha d'Água, an island situated in the bay. The pipeline broke about 2 km away from the refinery in a mangrove swamp next to the waters of the bay. The leakage continued for around five hours and the oil spread into the waters of the bay, where it formed a 50 km^2 oil slick. The cleaning and physical collection of the oil lasted for a month following the oil spill and at its peak it involved 2 400 workers. About 150 boats were used, as well as extensive amounts of barriers and absorbent materials. This oil spill illustrates the risks posed by high-risk industries to the neighbouring communities and the environment.

Risk perception frames

An analysis of the reasons for the oil spill was carried out. The analysis focused on the risk perception frames of the actors and how these frames differed from each other. Experts from the State Environmental Control Agency (Feema) associated the risk with the lack of environmental licenses for a hazardous operational system, which includes risk analyses for hazardous systems. Environmental licensing of hazardous operational systems has been obligatory in Brazil since 1981. The procedures include the licensing of new installations and the operational system, to be updated every five years. Risk analysis and external audits to evaluate the conditions of the installations, as well as emergency and contingency plans, are part of the licensing. Feema had for many years demanded the environmental licenses for REDUC but the requirements were never fulfilled. The conflict between Petrobras and Feema has been going on for several years:

> 'To fulfil the environmental licence requirement an external audit has to be carried out to evaluate the risk conditions in the refinery, but the external audit

> has for many years been denied by Petrobras. Instead, an internal audit has always been carried out by the experts of Petrobras' (President of Feema, *FSP* 20/01/2000).
>
> 'There are 33 units of Petrobras functioning around Guanabara Bay that do not have environmental licences' (Pres. of Feema, *Surgente*[1], n.1018, 19/01/2005).

On the other hand, Petrobras alleged that REDUC was established before the law was approved in 1981, and therefore the refinery was not obliged to fulfil its requirements. Many discussions were going on about an agreement that would permit external auditing at REDUC, but by the time of the 2000 oil spill, the audits had not been permitted by the Petrobras management.

The lack of environmental licenses has been seen by others as an inability of the state to fulfil its role as controller of environmental problems:

> 'How come that a company the size of Petrobras, with the complicity of the negligent state and federal authorities, has no environmental license and does not comply with the law? A Company that lets a spillage to go on for hours without any measures has no serious prevention programmes to respond adequately to its high level of environmental risks' (Senator from the Workers Party – PT, *O Globo* 26/01/2000).
>
> 'The media exclusively blamed Petrobras for the accident. However, government control agencies seem to consider that their role is exclusively to receive the fine to be paid after such accidents...Sindipetro understands that the environmental problems in Brazil are broader and more complex, starting with the posture of the federal government in relation to environmental issues' (Surgente 26/01/2000).

Other actors associated the risk of the oil spill on the fragile ecosystem of Guanabara Bay and the surrounding population with the inadequacy of the measures implemented by Petrobras to lessen the impacts of the spillage.

> 'The delay in taking decisions, the delay in communicating the accident to Feema, the lack of preventive measures, the lack of human and material resources to prevent the oil spreading made Feema suppose that Petrobras actually did not have an emergency plan prepared for the demands of major accidents' (President of Feema, *O Globo*, 24/01/2000).
>
> 'Petrobras was late with their emergency plan, so when it started it was too little too late. There were a lot of boats trying to help, but not all of them could get close to the oil. The oil was in a shallow area full of wood from old fishing traps, so it was hard for the boats to get there' (Leader of fishermen, in interview).

Petrobras agreed that there was a delay in detecting the accident, which contributed largely to the amount of oil spread into the water, which resulted in an uncontrolled situation. In addition, the natural conditions prevailing during the hours after the accident also contributed to the magnitude of the consequences of the oil spill (Petrobras top level manager in interview). This echoed with the

---

[1] Surgente is the newsletter of Sindipetro, the petroleum workers trade union.

words of the director of Petrobras, dismissed four days after the accident. He said that the emergency plan of REDUC was designed mainly for controlling smaller oil spills, at a maximum of 700 tons of oil (*FSP*, 26/01/2000).

However, other social actors, for example workers associations and trade unionists, related the risk of accidents with the risk management policy at Petrobras:

> 'The contracted workers are poorly trained, poorly assisted and poorly salaried ... At the end of a contract, the workers move from one company to another, creating no interactive relationship with Petrobras (President of Petrobras Engineering Association, www.comciencia.br, 12/09/2005).
>
> 'Degradation was implanted at Petrobras purposefully during this government to create an image of incompetence what would justify the process of privatisation ... Degradation culminated with the occurrence of serious accidents. Sindipetro denounced this already in 1997 after the accident in the same pipeline, but nothing was done to change the risky conditions' (*Primeira Mão*[2], 02/03/2000.

The workers institutions claimed that the authoritarianism of Petrobras had been responsible for the bad performance of the REDUC refinery in questions relating to safety, health and environmental management. They also claimed that REDUC had being unclear about its safety and health policies. The refinery had systematically forbidden the workers representatives to participate in accident investigations (Committee for Engineering and Architecture CREA/RJ Report 2000; *Surgente* no. 643, 2000).

Sindipetro also claimed that the increasing number of contracted workers was affecting the safety management of the company (Sindipetro, boletim 699). The lowest level of workers directly employed by Petrobras was reached in 2001. In 2002, for each worker directly employed by Petrobras there were three workers from sub-contracted companies.

Identity frames

An analysis of the political context of first 20 years of Petrobras suggests that the military government built a top-down administration in the company, with nuances of authoritarianism. During the years, an institutional identity frame was developed at Petrobras. This frame was probably influenced by the military administration, but also by the technical expertise being successfully developed within the company. The president of Brazil in charge in the period of 1995 to 2002 explained that the authoritarianism and secrecy at Petrobras was a legacy of the military administration in the country (Cardoso, 2006). The institutional identity frame was based on a posture of 'self-sufficiency' and the expertise held by the company was considered sufficient to solve all problems relating to the company. This institutional identity frame probably caused the lack of commu-

2 Primeira Mão is the Petroleum Workers Federation publication.

nication between the experts from Petrobras and other social actors, as for instance, government inspectors and workers representatives. Their identity frame of high competence influenced the way they saw other actors as less competent.

Feema predicted problems in the Petrobras installation, but on the other hand it lacked power to properly conduct its role of inspector. According to a senator from a political party opposed to the party in power:

> 'The lack of power of the state agency is more like a lack of will of the state government to impose inspection in a company that brings high incomes to the state of Rio de Janeiro' (Senator from Workers Party).

Most of the workers representatives were members of a political party. This political membership defined their identity, and influenced the building of a political identity frame. This identity frame defined their perception of risks of accidents, often in opposition to the federal government in charge at that time.

Conflict management frames

Lack of communication also existed between the Petrobras administration and workers representatives. Being part of such powerful company influenced the way the administration of Petrobras chose to deal with the conflicts: problem solving was to result from their own expertise and long-term successful experiences. For example, workers and their representatives were often banned from participating in accident investigations carried out by Petrobras (Surgente no. 1018, 19/01/2000). They claimed that the opportunity was never given to them to express their view, for instance in reporting risky conditions at the workplace, or in investigating accidents. Since 1997, when an oil spill occurred from the same pipeline as that in 2000, Sindipetro had predicted problems at REDUC. They had reported the possibility of a new accident at REDUC at the same pipeline, but they claimed that they were not heard by the Petrobras administration.

On the other hand, the discourses of the workers representatives indicated their preference for how the conflict should be handled; they wanted a change in the federal government as the only solution to Petrobras's mishandling. This appeal to a political action frame was clearly shown in their publications from 1995 to 2002. Very often the first page of the Sindipetro newsletter was stamped with the slogan OUT FHC. FHC are the initials of Fernando Henrique Cardoso, the president of Brazil between 1995 and 2002. The statement below also illustrates their preference:

> 'The catastrophe at Guanabara Bay is the responsibility of the political administration of the federal government' (Surgente, 19/01/2000).

Frame reflection

The frame analysis carried out within this case study illustrates what has been said in the literature: the importance of the identity and conflict management frames either in the intractability or resolution of conflicts. It was also seen how the different perceptions of risk may intensify the antagonism existing among the parties.

Throughout its fifty years, Petrobras has accumulated a great amount of experience and achieved technological excellence in offshore oil exploration. The steady increase in the volume of oil production over the years indicates this excellence. However, this technological excellence had not always been translated into the risk management programmes. This situation, however, has changed since the beginning of this decade. This fact is shown by a sharp decrease in the number of accidents and volume of oil spill (Figure A1). The data presented by Petrobras were confirmed in a number of interviews with trade unionists, fishermen and their representatives, environmental authorities and Petrobras employees. These sources of information indicate that there has been a change in the safety policy at the company as a whole, and at REDUC in particular.

> 'We though that we had certain environmental excellence, but the accident showed it was not sufficient ... I can affirm that we will take many lessons from this accident [referring to the accident in 2000]. The most important one is that we need to promote an environmental revolution at the company. Petrobras has developed a world-known technical excellence in deep waters; now it is essential to also develop environmental excellence. That is the only way to be internationally recognised... Up to now the environmental issues have not received the right attention, in consequence of lack of corporative vision in respect to environmental issues' (President of Petrobras, O Globo, 20/01/2000:13).

Fig. A1: Volume (m3) of spilled oil from Petrobras, 1997-2005.

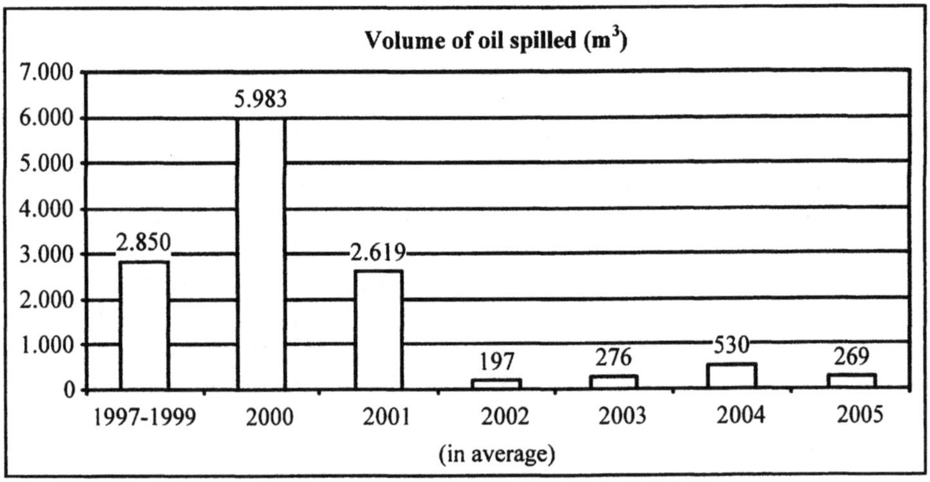

Source: *www.petrobras.com.br retrieved 12/10/2006*

> 'The turning point was the accident. After that, Feema and REDUC established a programme of adjustment of conduct as a consequence. Now they [REDUC managers] are more open to conversation and to establishing goals for operational safety and environmental issues' (Feema inspectors, in interview).

However, other social actors, for instance workers representatives and members of PT, see the change in safety culture of Petrobras as a consequence of the new federal administration that took office in 2003. In 2003, a representative of the Workers Party was for the first time elected the President of Brazil. They have the opinion that this government is keener on promoting occupational safety and health issues than previous regimes. Although there are indications that the change in the safety culture of the company started before this new administration, it is very likely that the new government had a positive influence on the relationship between the company and workers and their representatives.

The main purpose of this case study has been to investigate the usefulness of frame analysis in the study of environmental conflicts. The tools used to carry out frame analyses were: a) to place the conflict in a historical and socio-political context, b) to understand the physical and organisational settings that led to the conflicts; c) to identify interested parties, and d) to analyse the parties' positions within the conflict.

Notes on Sources

The information in this Appendix was derived from a number of sources, primarily from interviews with federal and state government officials and environmental inspectors, petroleum workers and their representatives, managers of the oil company, and fishermen and their leaders. Information was also obtained from organisational documents and archives, national statistics, historical and contemporary research, and newspaper articles collected from newspapers from the State of Rio de Janeiro and Sao Paulo. The time span covers the five years from the date of the accident (2000-2005).

Chapter 3

Case Studies

–

Loudden (Stockholm, Sweden)
Hamburg (Germany)
Haapsalu (Estonia)
Klaipėda (Lithuania)
Primorsk (Russia)
Ventspils (Latvia)

Loudden
–
A Controversial Harbour for Petroleum Products in Stockholm

Ronald Wennersten[1], Nils Brandt[1], Åsa Larsson[1]

Photo: Ronald Wennersten

1 Royal Institute of Technology (KTH) – Department of Industrial Ecology – Stockholm – Sweden.

Abstract

Stockholm has for long time been an important city for trade in northern Europe and the Baltic Sea Region (BSR). The scenario for the future is that this role will not be less important but the prediction is rather that the harbours in Stockholm will play an even more important role in the future. Loudden is one of several harbours in Stockholm County and the biggest oil terminal of the region. It is situated in the eastern part of the central Stockholm. After several years of infected debates between several stakeholders as politicians, oil companies, municipalities and NGOs, the city of Stockholm finally took a decision in 1999 not to prolong the contracts with the oil companies after 2011. Important motives for the decision to close Loudden were to get rid of the transportation of petroleum products by trucks travelling through the city and the transportation of petroleum products in the archipelago. Another important motive is also the city's strategy to build and expand the city inwards, which means to use as much of existing land as possible and in the same time use existing infrastructure for building residential areas and offices.

There are several stakeholders involved in the case. For example harbour company own by the city, oil companies renting the oil harbour from the city, neighbours to the industrial area, planners at city planning office, tourists, NGOs, political parties, etc. It's important to recognise that the interactions between stakeholders have consequences for the decisions that are made.

1. Introduction

Stockholm, sometimes called 'The Venice of the North', is built on 14 islands around one of Europe's largest and best-preserved mediaeval city centres. Stockholm was founded in the middle of the twelfth century at the point where Lake Mälaren, Sweden's third largest lake, meets the Baltic Sea. During the seventeenth century, Stockholm evolved into an important city of trade and shipping due to its strategic location. The Stockholm Archipelago is the biggest archipelago in Sweden, and one of the biggest archipelagos in the Baltic Sea. It consists of approximately 24,000 islands and islets.

2. Background

2.1 Description of Stockholm's Harbour and Loudden Dock

Stockholm is built on the waterfront, which means that the harbour has always played an important role for the city. Until the 1950s, Stockholm harbour was the biggest import harbour in Sweden. The city flourished and traded with all parts of the world. After the Second World War the trade with the Baltic States ceased, while ferries to Finland increased in numbers.

Today Stockholm harbour is the biggest harbour on the Swedish east coast and plays an important role in the provision of goods to the city and Mälardalen region and also for passenger ferries and cruising ships. The harbour is divided into several docks, the most important being Stadsgården, Frihamnen, Värtahamnen, Loudden and Masthamnen. Other docks of importance in the Stockholm area are the outports of Kapellskär (90 km north of Stockholm) and Nynäshamn (60 km south of Stockholm). In addition to Loudden, petroleum products are handled in the Södertälje harbour (40 km south of Stockholm) and the Berg oil terminal in Nacka (Malmsten 2005).

The increased demand for oil caused by industrialisation resulted in the construction of the oil port at Loudden in 1926. Loudden is the biggest oil terminal in the region, with an area of 250,000 m^2. It is situated in east-central Stockholm, close to the Stockholm National City Park (Fig. 1). It comprises more than 100 storage tanks plus six underground storage cisterns with a capacity of 130,000 m^3 for holding petrol, diesel, heating oil and other substances such as lubricants and naphtha. Furthermore, underground tanks situated in the south-eastern area of the Loudden docks are available but currently not in use. Loudden is thus an important component of the regional supply of petroleum products to central parts of the Stockholm Region and for shipping traffic.

Photo 1: The Loudden oil terminal in central Stockholm (Ronald Wennersten)

The organisational structure of the parent company Stockholm Harbours (Stockholms Hamnar) is presented in Figure 2. Stockholm Harbours is a subsidiary company and fully controlled by Stockholm city. The land is owned by the City of Stockholm.

Fig. 1: Organisational diagram of Stockholms Hamnar AB

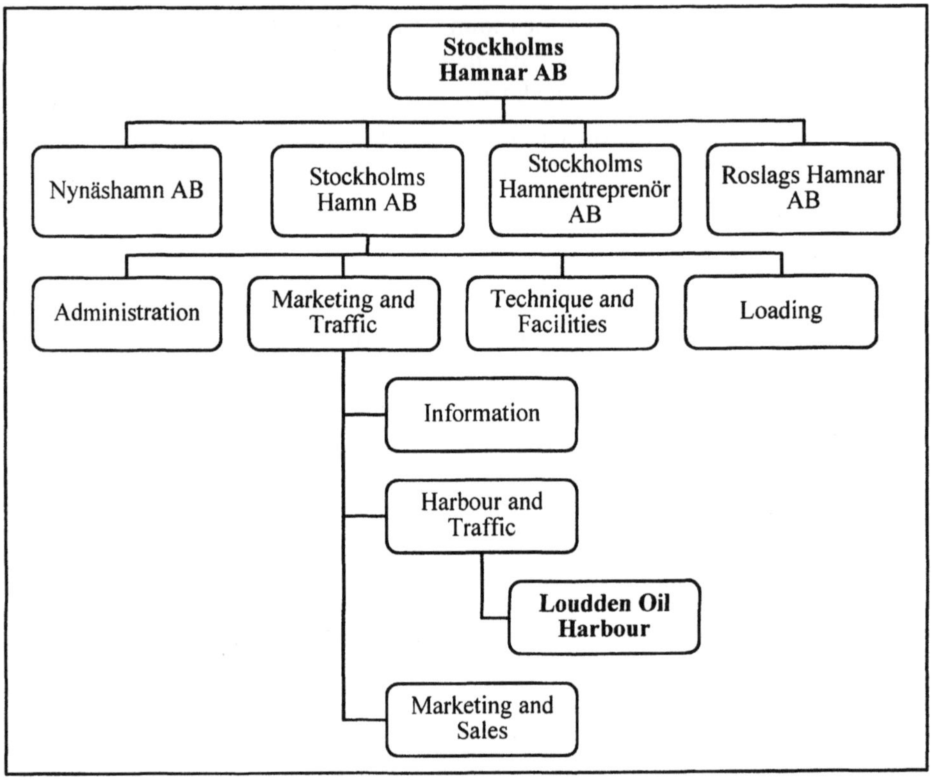

Source: Ronald Wennersten

The docks in central Stockholm, Kapellskär (north of Stockholm) and Nynäshamn (south of Stockholm) are important links in the regional supply chain, as well as for the sea traffic over the Baltic Sea.

There are nine oil companies active at Loudden, two decontamination companies and one bunkering company situated in the area, including:

- Preem Petroleum AB,
- Norsk Hydro Olje AB,
- Univar AB,
- Petrolia AB,
- OK-Q8 AB,
- AB Svenska Shell,
- Reci Industri AB.

Except for areas rented by the operating companies, Stockholms Hamn AB maintains common facilities such as piers, traffic routes and the gateway. Stockholms Hamn AB is responsible for safety and environmental reports on the harbour area.

Petroleum products such as diesel, gasoline and heating oil come in by ship from refineries in Finland, Norway, Sweden or Denmark through the archipelago, are unloaded and then stored in the Loudden cisterns and underground tanks. These 1.2 million tons of petroleum products and 170,000 tons of coal are mainly distributed by road traffic in the Stockholm Region.

The Swedish Maritime Administration (Sjöfartsverket) has stipulated that the capacity and measures of oil tankers bound for Loudden should not exceed 20,000 tons, 11 m draught and 200 m visibility in daylight. Safety factors reduce these values for night-time or bad weather conditions. All ships handled at Loudden in 2002 were equipped with double hulls. This is not due to legal requirements, but to the introduction of financial bonuses for ships with higher safety and environmental standards. Sheet pile walls have been built along the coastline of the harbour to prevent uncontrolled oil discharges from contaminating the land. Oil traps and sand filters have been installed to clean the rainwater run-off.

After several years of intensive debate between stakeholders such as politicians, oil companies, municipalities and NGOs, the city of Stockholm finally took a unilateral decision in 1999 not to prolong the contracts with the oil companies at Loudden after 2011. There was political consensus regarding this decision but no alternative site for the oil terminal has yet been identified. A solution for the region will affect not only the city of Stockholm, but also several other municipalities in the region. It is of course of central importance that a sustainable alternative solution to Loudden is presented and accepted by the stakeholders before Loudden is closed in 2011.

2.2 The case study around Loudden Docks

This case study was chosen because of its complexity regarding the decision-making process and stakeholder interest. The case study was mainly carried out in collaboration between the Royal Institute of Technology (KTH) in Stockholm and the City of Stockholm. Photo 2 shows Loudden Docks along the waterfront, almost embedded in National City Park which stretches through a large part of Stockholm. The area where Loudden Docks is situated is very close to the city centre and the Royal Institute of Technology (KTH). This location makes Loudden ideal for many purposes, such as a distribution node for petroleum products to the city but also potentially as residential area, recreation area etc.

Photo 2: Location of Loudden Docks on the Stockholm waterfront

Photo: *Ronald Wennersten*

The objectives of the case study were to:

- Study the decision-making process.
- Determine the knowledge, information and data that should form the basis for a decision.
- Reveal the interests and conflicts surrounding the process.
- Investigate the role a university can play in the mediation process.
- Identify conflict resolution methods that could be used in the process.

It is important to understand that the decisions taken in the administrative and political processes of a city are not only the result of information and data. Even if the situation concerning information is effective and all stakeholders have access to the same knowledge and information, different stakeholders will have various opinions and will favour different decisions. The important thing is that the stakeholders have trust in each other. The degree to which one party trusts another is a measure of belief in the benevolence and competence of the other party. A failure in trust can be forgiven more easily if it is interpreted as a failure of competence rather than a lack of benevolence. Knowledge, information and data should form the basis for a decision but the decision can only be taken in a process where the stakeholders' values are revealed. We only know of one such effective process, and that is the democratic process. The problem in practice with the democratic process is the real or experienced differences in power between the stakeholders.

Experience from many cases where conflicts or potential conflicts have been handled in different ways show that it is always important to start early in a conflict resolution process. The money and time spent on handling problems before they turn into intractable conflicts provide a very high rate of return on investment. It is not always possible to reach consensus, but the important thing is that different opinions can be made visible and it is always much easier to establish a trustful communication process early. It is also advantageous if a third party can play an important role in the communication process. This third party should have an independent role in the process. From investigations of public confidence in different institutions, we know that universities are always perceived with a high degree of confidence, while politicians and industry are poorly rated. Universities should therefore strive to establish value-free platforms for discussions and develop scientifically-based evaluation tools.

One of the objectives in the present case study was thus to see how a university (KTH) could play the role of mediator in resolving conflicts such as that surrounding Loudden. This was achieved through the use of different methods, examples of which are:

- Creating an information base through a literature study: Reports, books and newspapers were used to collect information needed in the case study.
- Workshops and hearings: Workshops and one hearing were used to create platforms for discussion between different alternatives for Loudden and also around the decision-making process.
- Indicators were developed describing different aspects of sustainable development for different alternative locations of the docks.
- Risk assessment: The aim here was to develop methods to assess risks so that the results could be communicated to all stakeholders in the process. The methods had to be based on knowledge from research in risk communication.
- Questionnaire: A questionnaire survey was carried out in order to obtain information on how different stakeholders experienced the decision-making process and what changes they would have liked to make to it.

2.3 Administrative, political and legal framework

It is not possible to fully understand the decision-making process around Loudden Docks without understanding the administrative, political and legal framework in Sweden. The location of docks in Stockholm is part of an integrated coastal zone management (ICZM) process for the region. However, there is no overall national legislation or special agency or institution responsible for planning, implementation and evaluation of ICZM in Sweden. The main legal framework is the Swedish Environmental Code and the Planning and Building Act, which apply to both terrestrial and marine areas. The inland and off-shore extent of the coastal zone is not explicitly defined in the Code but the areas were comprehensively shown on maps connected with the Bill in which the regulations were proposed. According to the Bill, the more detailed definition of the various coastal zones is a task for municipal comprehensive (physical) planning. The extension off-shore is suggested to be 1-3 nautical miles from the baseline. A major part of the coastal zone has been identified as an area of national interest. Many of the provisions in the Environmental Code must be used when applying the Planning and Building Act. The special management provisions in the Environmental Code include regulations on sitting new industrial installations, tourism and recreational functions and restrictions on summer cottage developments.

The central government ministries most relevant to coastal zone planning and management are the Ministry of the Environment and to some extent the Ministry of Agriculture, where issues concerning fisheries are handled. The National Boards and Agencies in Sweden also have great responsibilities regarding environmental issues and planning, since they produce guidelines and since some of them give permits to larger projects within their sector.

Most important are:

- The Swedish Environmental Protection Agency (EPA),
- The National Board of Fisheries,
- The National Board of Shipping and Administration,
- The Swedish Board of Housing, Building and Planning,
- The Planning Coast Guard.

The municipalities in Sweden are responsible for physical planning at both the comprehensive and detailed level, although the County Administration Board can intervene if decisions by the municipalities threaten national interests.

The Swedish government and some regional organisations support sustainable development in coastal regions. The Regional Environment and Management programmes are examples of the ambition to integrate ecological, economic and social aspects in regional development.

The immediate power of planning in Sweden lies with the local municipalities. There is a long and strong tradition in Sweden of local self-government, and the municipalities are strong compared to the situation in many other countries. However the government and its regional agency the County Administration Board have to monitor all planning and have the power to object in certain cases – mainly when national interests, health and security and inter-municipal interests are violated.

In contrast to the case in many other countries, the Government departments in Sweden are relatively small. The Ministry of Environment – responsible for the environment, planning and building – consists of 125 members of staff.

To support the government administration, there are a number of central authorities or boards, each responsible for a sector of the community. To support the Ministry of Environment there are six different major central authorities employing approximately 1,500 people. The Government can give the authorities instructions concerning their policies and activities, but according to the Swedish constitution the Government is not allowed to steer their decisions in individual cases. In this respect the authorities are independent of the Government. This is typical for the Swedish constitution, which dates back to the 17th century.

The Swedish National Board of Housing, Building and Planning, with 165 employees, is one of these major central authorities.

On the regional level, Sweden is subdivided into 21 counties headed by county administrations linked to central authorities. The County Administration Board is a state agency operating under general directives issued by Parliament and the Government. The board is completely independent to take decisions within its own framework. In some cases the County Administration acts as a court. The head of the county is a Governor, appointed by the Government for a period of six years. The Governor is chairman of a board of 14 people, to which the members are indirectly elected.

Parallel to the County Administrative Board, which is a government agency, there are on the regional level County Councils. These are elected by the people in general elections every four years. Their main purpose is to be responsible for healthcare. In Stockholm, the County Council is also responsible for regional planning and for public transport. The Regional Planning Office, which is an agency under the County Council with its own political board appointed by the Council, works with the regional planning issues. The company *Storstockholms Lokaltrafik* is owned by the County Council and operates the public transport network within the county.

The smallest political body is the municipality. Sweden has 288 municipalities, with populations varying in size from 3,000 to 750,000. The median size of a Swedish municipality is approx. 16 000 inhabitants and the median area approx. 700 km^2. Stockholm City is the biggest municipality of all, with 750,000 inhabitants.

The supreme decision-making body in a municipality is the Municipal Council, in the case of Stockholm the City Council. It is directly elected every four years and has in the case of Stockholm 101 seats. The Council appoints a number of delegations with about 15 local politicians, some of whom also have a seat on the Council. The most important delegation is the *Executive Board.* Another important and compulsory board is the *City Planning Committee.* In Stockholm other important boards are the *Real Estate and Traffic Committee* and the *Environment and Health Committee.* These committees have offices at their service, most interesting here are *The City Planning Administration, The Traffic and Street Cleaning Administration* , *Stockholm City Development Administration* and *the Environmental Administration.*

Stockholm City owns the *Stockholm Harbour Company,* which operates the docks in Stockholm, Kapellskär (in the north-eastern part of the county) and Nynäshamn (in the southern part of the county). Loudden is located within the Stockholm Harbour area.

3. Description of the conflict and the decision-making process

The conflicts around the Stockholm docks are multidimensional and can be divided into three main areas:

- Protection of the environment and public health.
- Social aspects such as availability of residential apartments, job opportunities.
- Economic aspects mainly for municipalities and companies.

Examples of concrete conflicts within these areas are:

- The future need for docks: This will be dependent on the development of the City of Stockholm, as well as the future demand for fuels.
- Economic interest from other stakeholders: The companies that are active at Loudden have expressed their interest in staying at Loudden and even expanding some activities there. Other municipalities in the region have expressed their interest in taking over capacity from Loudden, mainly for reasons of economic development in their area.
- The future need for apartments and offices in central Stockholm: Part of the policy is to expand Stockholm inwards the area where the Loudden Docks is situated. It would be very attractive for the city of Stockholm to build residential areas and offices.
- The protection of the Stockholm Eco Park, defended by several NGOs.

In 1999, Stockholm municipality decided that the Loudden oil terminal should be closed down in 2011. One motivation for the decision was to get rid of the transportation of petroleum products by trucks travelling through the city and the transportation of petroleum products in the archipelago. Before the City Council took the decision, there had been discussions around this issue for many years. Initially, the discussion focused on the supply of aviation fuel from Loudden to the Stockholm airports at Arlanda and Bromma. During the 1980s discussions were held about a pipeline as an alternative means of transportation of the fuel. In 1994 the City of Stockholm decided to guarantee the funding for such a pipeline. One of the conditions for such a guarantee was that the company responsible for the pipeline would have long-term contracts with the companies delivering aviation fuel to Arlanda. However the oil companies did not want to commit themselves to such a long-term agreement and this ended the discussions around the pipeline. This is a basic problem connected to right of disposition, where the city is often responsible for planning infrastructure but it does not have any control over the companies delivering products, e.g. petroleum products and energy (Lidquist 2005; Malmsten 2005).

The decision to close the Loudden Docks gradually developed up to 1998, when an accident occurred close to Loudden. The safety aspects can often radically change public opinion when an accident occurs. This happened when a truck containing 14 tonnes of propane was involved in an accident close to Loudden, in the central part of Stockholm in 1999. This accident did not lead to any severe consequences, but it could easily have caused a catastrophe. It had a great impact on the discussions about the risks associated with transport and led to a ban on the transport of propane from the harbour. One of the parties in the City Council, the Stockholm Party, used the accident for one of the matters close to its heart, namely to get rid of heavy transport from Loudden through the city and through the archipelago. The Stockholm Party formed an alliance with the non-Socialist parties to run the city, and the City Council of Stockholm took the decision to close the Loudden Docks. When it came to signing new contracts with the oil companies, there was a new strategy. The previous contracts ran for 20 years in order to give the petroleum companies possibilities for long-term investment, but from now on the contracts were only for one year. In 2002 there was a new election when the power went over to the socialist parties and the environmental party. The new City Council opted to sign contracts with the oil companies up to 2008, with the possibility for prolongation to 2011. (Lidquist 2005).

The decision to close the Loudden Docks after 2011 has not been changed and there is now a process running for finding a new solution for the regional supply of petroleum products. During autumn 2006, the handling of aviation fuel was moved to the harbour in Gävle, a town 175 km north of Stockholm. Fuel is taken to the docks in Gävle and carried by railway to depots at Brista power station from where it is pumped through a pipeline to a depot at Arlanda airport.

Another important motive for closing the Loudden Docks was Stockholm City's comprehensive strategy to transform industrial land to build new housing. The Stockholm City Planning Committee is planning to build a new residential area on the property where Loudden is located. In its comments on the decision of the City to close the Loudden Docks, the County Administrative Board has written:

> „The docks are of strategic importance for the region as the only full service docks and are in this role of national interest. How the docks' activities will develop in the long run is unclear. There is however much that indicates that the area will continue to be docks with surrounding activities in the foreseeable future. The area has an attractive situation in the city and can, when the docking activities decrease, be converted to a city development area. The possible phase-out of the docks must however be seen as a long term development"

Today the County Administrative Board of Stockholm is searching for an alternative solution for the petroleum docks in the Stockholm region. However, there are many questions that need to be addressed and many interests to be handled before a solution can possibly be accepted by all the stakeholders.

Some of these are:

- The strategic supply of petroleum products for the region.
  How will the demand for the products develop in the next 10-20 years? Stockholm City has established the goal to be fossil fuel free by 2050 (Environmental and Real Estate Division, Stockholm City Council 2007). Do we need docks for petroleum products in the long run? How will a gradual switch to bio fuels affect the situation?
- Interregional competition between the municipalities for employment opportunities, infrastructure development and economic growth. Docks would mean employment opportunities and economic growth for a municipality, but would also require adjustments of the infrastructure to support transport to and from the docks. What development is necessary and who will pay?
- Which aspects in sustainable development of the regions are most important? Environmental impacts, risks, social aspects, culture aspects?
- Who will bear the costs for phasing out the docks, decontamination etc?
- Who is responsible for finding an acceptable solution for the oil companies?
- Their interest is that a change will not raise their costs.
- What should the land at Loudden be used for if the docks are closed? What consequences will a new residential area have for other activities in the area such as ferries etc? The area is also close to the National City Park.
- Who should take the final decision and what impact should the different stakeholders have on the final decision?

Many stakeholders are affected by the decisions but the actual decisions are always linked to the actual power to decide. In the planning process it is important to have an early exchange of thoughts around different solutions among the stakeholders. This can be classified as upstream conflict resolution. Before the decision was taken to close the Loudden Docks by Stockholm City Council, there had been a long period of discussion among the stakeholders. However these discussions had been diffuse and unstructured in many ways and the decision was finally taken with the Stockholm municipal interests in focus. The main problem connected with the decision was that no alternative for location of the docks had been decided. The decision was taken by Stockholm City Council as one of the stakeholders, but the one holding the decision-making power. However, the alternative locations will affect the other stakeholders, such as the petroleum companies and the other municipalities in the region. Thus the decision was not taken as a balanced decision among the different conflict aspects and regional interests. The main problem with this kind of decision is that it tends to create intractable downstream conflicts, as in the case of Loudden.

For this reason it is important to look for conflict resolution methods that can be used to reveal information and values among the stakeholders as early as possible in the planning process. It is of course also important to identify at an early stage the stakeholders who will be affected by later decisions and plans.

4. Relevant stakeholders and relevant conflict resolution methods

4.1 Stakeholders involved in Loudden Docks relocation process

In Table 1, the stakeholders identified in the conflict are categorised into interest groups, individuals, organisations and authorities.

Tab. 1: Name and nature of stakeholders involved in the Loudden Docks relocation issue

Name of the stakeholder group	Interest groups	Companies	Organisations	Individuals	Authorities
Stockholms Hamn AB		x			
Oil companies	x	x			
Neighbours to industrial area				x	
Tourists			x	x	
Citizens	x			x	
Consumers	x		x		x
Swedish Government					x
Political Parties			x		

Source: Ronald Wennersten

Stockholm's Hamn AB can be considered one of the most active groups in this conflict. This might be due to the fact that on the one hand the chairman of this company is actively driving the question of shutting down the oil terminal at Loudden and on the other hand the civil servants, managers and workers in the company do not want to move the facility. Many internal investigations about possible alternatives, risks and environmental impacts have been conducted and are presented in this paper. Stockholm's Hamn AB is part of the stakeholder group.

The *oil companies* (generic term for all companies operating at Loudden) are acting in a more defensive way. A few investigations have been carried out by e.g. OK/Q8 regarding the risk assessment of transportation of petroleum products. Due to increased occurrence of oil spill accidents in the Baltic Sea, the public awareness about the problem is omnipresent and the companies have to work on their environmental image.

On the other side there are the attempts of Preem Petroleum AB to increase the volume of petroleum products allowed to be handled at the Loudden oil terminal (from today's 229,000 up to 818,000 tonnes per year). This case is currently being discussed at the Environmental Court in Stockholm. Because of these arguments, the oil companies form one group of stakeholders in the target group.

Of course *neighbours* to industrial areas in general would preferably not live close to such a site, especially in the centre of the city. The City Council of Stockholm is aware of this problem and is therefore declaring parts of the city to be environmentally protected areas, e.g. the National City Park in the direct neighbourhood of the Loudden oil terminal.

Neighbours are complaining about noise, smell and air pollution by traffic and would like the oil terminal to be moved to another place. People living along the access roads are also disturbed by trucks passing by and polluting the environment with noise and exhaust fumes and the residents are afraid of being seriously injured in the event of an accident occurring (explosion, fire etc.).

In addition, the level of sensitivity has been increasing in recent decades: people know about the health effects of 'dirty industries' and want to avoid harmful substances. On the other hand practically everybody is dependent on petroleum products: for example driving cars or using public transportation, using heating systems and electricity, as well as travelling by boat or plane.

Coming to the question of whether or not neighbours should be a part of the target group, the decision is based upon the assumption that political parties and organisations, such as the Organisation for Environment and Social Development, are generally represent the points of view of social groups such as neighbours and citizens, especially since no civil action groups have been established.

Tourists coming to the City of Stockholm by ship from other countries around the Baltic Sea land in the direct neighbourhood of the Loudden oil terminal. Other tourists will probably not come into contact with this area since no sightseeing attractions other than the National City park and the Kaknäs Tornet, giving the view from above, are in the close surroundings of Loudden.

It is not very likely that the impression of Stockholm and Sweden would dramatically change from a tourist's point of view, if Loudden were to be moved in the future, but the City would definitely be able to create a new tourist magnet by building a new quarter with shops and museums.

The interests of *citizens* and neighbours in a democratic state are presumed to be represented by political parties or organisations, as long as no action groups are formed.

Within the group *consumers*, members of other groups are partly included, such as citizens using cars or public transport, neighbours using oil for their heating systems and also other industrial companies that are using these petroleum products in further production steps. These consumers are interested in this discussion because it might have an influence on the market prices for petroleum products in the event of the oil terminal being moved to a more distant location.

The *Swedish Government* is the stakeholder that aims to look at the national economic, environmental and social aspects of this conflict. This perspective allows and requires a thorough evaluation of the conflict. Even if there are lifelike different points of view towards this problem in different political parties, the government as an institution has certain regulations and guidelines. Because of this, the government is part of the target group as a representative of citizens and neighbours.

4.2 What type of information is needed in such a complex decision?

As has been described, the decision-making process around Loudden is very complex, involving many stakeholders and conflicting interests. To facilitate an effective decision-making process, there is a need for a comprehensive background material based on a holistic view and if possible a common vision or visions. The vision should be an integrated part of a vision for sustainable development of the Stockholm region. This is also a central goal for the planning process in the City of Stockholm, to focus on sustainable development. Vision Stockholm 2030 states (City Planning and Traffic Division, Stockholm City Council 2007):

> „The City of Stockholm is working on a holistic and long-term vision in the project 'Vision Stockholm 2030'. The vision shall serve as inspiration and uniting force in the long-term development carried out in the City of Stockholm and other actors in the region. The work is founded on a broad cooperation and agreement around strategic questions for the Stockholm region."

In developing this vision the city strives to engage all stakeholders in the city in an active process of discussion. Sustainable development can be formulated in a general way, such as that in the Brundtland report from 1987: „development which meets the needs of the present without compromising the ability of future generations to meet their own needs."

When it comes to practice, this general statement has to be broken down into more concrete objectives and goals. A vision has to develop some targets and goals and also indicators that will show how progress is achieved. Indicators can be formulated top-down as in the case of the Swedish Environmental Goals, which are broken down into different sector goals e.g. for energy and building sectors (Swedish Environmental Protection Agency 2007).

However, sustainable development is a complex issue involving choices and values about what is more important or less important. These choices can only be made in a democratic process involving broad public participation. In this sense it is also important to formulate visions to develop indicators in a bottom-up process.

In creating the vision of a sustainable Stockholm region, there are several key components central for the development of the region in relation to the case study. Examples of these are:

- National transport political goals.
  The Swedish government decided in 1998 around a general goal for transport policies in Sweden. The overall goal was formulated as 'To ensure a national economically effective and sustainable transport support for citizens and business in the whole country'.
- Scenario(s) for the development of the energy needs in Stockholm County.
  In a recently published report, a prognosis for the future consumption of petroleum products in the Stockholm region was described. Stockholm County has the fastest increase in population in Sweden. The prognosis shows that the consumption of petrol and diesel in the Stockholm area will increase for the next 20 years. The consumption of diesel is expected to increase the most (Transek 2005).
- Scenario(s) for how these needs are going to be met.
  A change from fossil fuels to alternative fuels does not mean that docks will be redundant, since for example most of the ethanol used in Sweden is imported from Brazil. However the alternative fuels have a lower energy value and greater storage volumes are needed. A change to alternative fuels would thus require larger storage space and larger transportation capacity.
- Transport and depots – the planning of infrastructure for the coming 20-30 years.
  Stockholm is a large region, with 40% of the petroleum products being consumed in the central part and the remainder divided between the north and south of the region. Several investigations propose that docks located north and south of the Stockholm region should replace Loudden, to avoid heavy oil transportation through Stockholm city centre. Several of the suggested harbour locations are situated far from Stockholm and calculations show that transport distances will increase by 25% if Loudden is closed (Transek 2006).
- Regional and municipal general plans.
  In the Stockholm region there is a regional plan called RUFS which describes the region's strengths and weaknesses. The plan is a strategic instrument to integrate continuing growth with long-term sustainable development. All communities in Sweden must have a comprehensive plan that describes how land and water resources should be used in the community.

In addition to these key components, the background material should include:

1. Environmental consequences
   - Emissions to air
     At the docks, the petroleum products give rise to volatile hydrocarbon emissions when the petroleum products are pumped to the cisterns from the ships. When oil is transported by ships and trucks, the transport facilities generate emissions to air and the oil give rise to emissions during unloading. The emissions from the ships contain sulphur oxide, nitrogen oxides, carbon dioxide and particles. The consequences can be acidification, eutrophication, trophospheric ozone, greenhouse effects and depletion of the ozone layer.
   - Emissions to water
     Oil spillages can occur at different places in a dock area, for example at access ramps for trucks and railways, quays, pumps and tank locks. Contaminated water can also be released from the underground storage cisterns. The consequences of oil spills can be poisoned drinking water, contaminated nature and eutrophication.
   - Waste
     Waste in an oil terminal mainly consists of sludge from the oil sludge separation, cleaning residues from tanks, oil polluted matter from the ground, waste from the oil tanker engine house, flushing water and ballast water.
   - Protected areas
     It is important that sensitive areas and objects close to docks, depots, channels, roads and railways are mapped to determine how sensitive these areas are during standard operation and transportation and what the consequences would be if an accident occurred.
   - Erosion
     Erosion at coastlines, beaches and sea beds can be natural or generated by traffic. Erosion generated by ships depends on the speed, size and shape of the ship, the gradient of the beaches, the width and depth of the channel and the quality of the sea bed.
   - Noise
     Activities in harbours and depots can create noise that disturbs residential and recreation areas in the local environment. Noise from sea traffic is mainly generated when the ships are passing through the channel, lying in the quay and when the oil is being loaded and unloaded at the docks.
   - Vibrations
     Intensive transport by heavy traffic, e.g trucks and railway, creates vibrations in the land and poses a high risk to buildings and nature.

2. Risk aspects

   The risks in docks are mostly associated with handling of the oil products. The products are classified as highly inflammable or less inflammable. The products can catch fire if leaks and evaporation occur.

   Some docks and depots are located in industrial areas where other businesses can intensify the consequences if an accident happens in the oil terminal.

3. Transport
   - Road transport
     Most accidents involving transportation of dangerous goods in Sweden affect nature and water resources, and the consequences are serious but fatalities are unusual. To make transportation safer, it is important to map dangerous road sections; bridges, tunnels, sections with very intense traffic and sections recommended for dangerous goods transportation.
   - Railway transport
     Railway accidents occur in different ways, for example level crossing accidents, sabotage and derailment. The standard of railways for dangerous goods transportation is important, for example number of level crossings and tunnels, how accidents affects roads and other railway lines nearby and whether different products should be transported in the same train.
   - Sea transport
     Discharge from oil tankers during normal operation consists of contaminated ballast water, cleaning water from the tanks, oil residues and water from the engine room and the keel. Grounding and collisions can lead to accidents with larger disasters. In the Baltic Sea, grounding is the most common accident and collision the next most common. Oil deliveries are limited by the quality of the harbour channels, quay berth access and road and railway capacity. The Swedish maritime administration is responsible for drawing up regulations regarding the types of ships permitted to traffic Swedish channels, depending on the width, depth and visibility in the channels.
   - Transport costs
     The price varies between different means of transportation for petroleum products. For example the price is 0.05 SEK /ton km for transportation with oil tanker and 0.10 SEK/ton km for transportation by rail in both directions.

4. The present standard and capacities of the harbours in the region

   When Loudden is closed, the Loudden national interest must be abolished and the environmental permits at the alternative docks must be consulted. The Swedish maritime administration decides which harbours and channels should be classified as national interest. The environmental permits regulate how much oil products each harbour is allowed to manage.

5. Costs of reconstruction and decontamination of Loudden when the docks close
   - Land and large investments are required to construct a new depot. In investigations concerning Loudden, calculations show that a new depot (500,000 m^3/year) would cost approx 300 mil. SEK and an inland depot (500,000 m^3/year) would cost approx 300 mil. SEK (Transek 2006).
   - Activities in the Loudden Docks have been going on for 80 years and the ground in the area is contaminated. Decontamination of the area can only begin when all activities that can cause contamination have ceased. The business companies in the Loudden Docks area are responsible for the decontamination. One of the specific problems with remediation of the area is that there are several bedrock storage systems. Remediation of these could require a long time.

As can be seen from this short summary, there is a need for an extensive background material in order to describe the complex issue around the Loudden case. Such a background material has to be in such a form that it can be communicated to the stakeholders in the decision-making process.

The decision-making process for this kind of regional problem is unclear. There is no regional forum for such processes and the decision taken by the City of Stockholm is a unilateral standpoint although many regional stakeholders are affected.

4.3 What methods can be used to facilitate the decision-making process and create platforms for discussions?

One of the central objectives in this project was to investigate how different methods can be used to create platforms for mediation between the stakeholders in the conflict. The outcome of the process is highly dependent on the willingness and commitment of the stakeholders to find a reasonable and practical solution. The parties have been involved in the processes and conservative in their opinions for a long time, and it could be necessary to involve a mediator in order to create a basis for a fruitful working environment. It is necessary that the stakeholders agree on such conflict assessment beforehand.

In this case study, different methods for conflict resolution were applied. The most important methods used were:

- Hearings, seminars and workshops,
- Questionnaire survey,
- Risk assessment,
- Indicators for sustainable development.

4.3.1 Hearings, seminars and workshops

Hearings are often used as part of a juridical process but can also be used as part of conflict resolution strategy where different stakeholders have the possibility to ask a committee questions. In this case, a hearing was held with the politicians in the municipal district committees and an energy company planning for a large storage facility of natural gas close to the Loudden area. The hearing was directed by KTH, which took the role of a neutral third party, and the focus in the hearing was safety issues. In order to get permission to construct the storage facility, the company had to include a written safety report in the permission application which was in a standard form. However these safety reports are very difficult to understand and evaluate for laymen and had been criticised by a local NGO.

The hearing was held with around 15 participants without the obligation for any party to undertake any commitments. The aim was to create a more neutral platform for unbiased discussions in the presence of neutral experts from the university. The results of the hearing were very positive and a step forward in the communication process and in understanding the stakeholders' arguments. An interesting result was that the role of the university as a third party was very appreciated by the participants in the hearing.

In order to collect information for the case study, two workshops were held. During these workshops different stakeholders had the possibility to give their opinions on the questions around a new harbour structure and also to comment on other stakeholders' opinions. The workshops were held by KTH, acting as the convener and also as the moderator creating a neutral ground for the discussions. In order to further develop the contacts with the stakeholders in the case study, a student has been working on her Master's thesis in parallel, on the theme 'Sustainable harbour infrastructures for the Stockholm region'. This work has created more regular contacts with the stakeholders during the case study.

4.3.2 Questionnaires

Questionnaires are a type of statistical survey handed out to a specific group of persons to gather information. Questionnaires have advantages over some types of surveys in that they are cheap, do not require as much effort from the questioner as verbal or telephone surveys and often have standardised answers that make it simple to compile the data.

In order to better understand how the different stakeholders viewed the decision-making process, in this case study a web-based questionnaire was sent out in 2006. The questionnaire was aimed at administrative (executive) officials within different stakeholder groups such as the County Administration, County Council, different municipalities, docks, oil companies and NGOs. The questionnaire was answered by around 70 officials (of approx. 200) mainly from county administration boards and municipalities. The results created an important discussion among the stakeholders around the need for a regional body handling these kinds of planning processes. One interesting result was that information from newspapers seemed to be one of the most important information sources for the officials. Another interesting result from the questionnaire was how the officials prioritised different aspects when planning a new structure for the docks area. The results showed that the officials gave transport opportunities to and from the docks together with environmental concerns as the highest priority, while employment was ranked lowest.

The results from the questionnaire also showed that it is mainly stakeholders from Stockholm that are taking part in the docks planning process and that these stakeholders think that the process affects a bigger region and that more stakeholders outside Stockholm should be active in the process.

Today there is no regional political organ that works together for finding a new solution for the docks. Of those stakeholders that responded to the questionnaire, 33% believed that a regional organ should take the final decision about how the oil deliveries for Stockholm region should be organised in the future when Loudden is closed.

The stakeholders want to replace Loudden with several docks located south and north of Stockholm to avoid long-distance oil deliveries by trucks through the city. In the planning process for the new docks structure, the most important factors to consider are the transport possibilities by sea, road and railway, environmental impacts, risks and regional development.

In the future, the stakeholders believe that the function of the oil terminal will change and other fuels will be delivered to the docks, for example biofuels. On the other hand, however, the majority of the stakeholders believe that the Stockholm region oil supply will always be an important question.

In 2006 a seminar was arranged by KTH and the results from the questionnaire were presented. Some topics in the questionnaire were discussed in more detail during the seminar, for example who should make the decisions about the future docks solution for oil supply in the region? Will we be using fossil fuel in Stockholm in 2050? At the seminar the stakeholders agreed that a special regional organ is needed in the docks planning process to deal with issues concerning land use, infrastructure and energy consumption.

4.3.3 Risk assessment

In 2006 a seminar was arranged by KTH and the results from the questionnaire were presented. Some topics in the questionnaire were discussed in more detail during the seminar, for example who should make the decisions about the future docks solution for oil supply in the region? Will we be using fossil fuel in Stockholm in 2050? At the seminar the stakeholders agreed that a special regional organ is needed in the docks planning process to deal with issues concerning land use, infrastructure and energy consumption.

- Describe real Worst Case Scenarios, not more probable but less catastrophic scenarios. Realise that calculations of probabilities are very unreliable. Try to evaluate your ability to handle Worst Case Scenarios.
- Do not underestimate the consequences in content and discussions. If one party focuses on probabilities, other interest groups will focus on consequences.

4.3.4 Indicators

The DPSIR (Driving force; Pressure; State; Impact; Response) framework is not a set of indicators but rather an approach often referred to in the context of sustainable development indicators. The concept underlying the DPSIR framework is cyclical: human activity exerts Pressures on the environment resulting in changes in its State; such changes will have an Impact on human and ecosystem health, which in turn may elicit a Response for corrective action and changing habits. In this case study, the DPSIR model was used to develop indicators for use in the discussions around the sustainability of different alternatives for dock relocation. From the set of indicators developed, a subset can be chosen in an evaluation process where the most relevant indicators for the docks structure are chosen. In earlier investigations for sustainable docks structure, the focus was more on the national economy and less on environmental consequences (Bosch & Gabrielsen 2003; European Swedish Environmental Protection Agency 1999).

In the planning for a new docks structure, a number of issues have to be addressed, for example emissions, demand for oil products, political decisions etc. With the DPSIR model, both qualitative and quantitative indicators can be developed and the connections between them can also be visualised in order to decide which actions have to be taken in order to establish a more sustainable docks infrastructure.

5. Conclusions

In the planning for a new docks structure, a number of issues have to be addressed, for example emissions, demand for oil products, political decisions etc. With the DPSIR model, both qualitative and quantitative indicators can be developed and the connections between them can also be visualised in order to decide which actions have to be taken in order to establish a more sustainable docks infrastructure.

6. References

Bosch, P. & Gabrielsen, P. (2003). Environmental indicators: Typology and use in reporting. European Environment Agency.

City Planning and Traffic Division, Stockholm City Council (2007). Vision Stockholm 2030 – Underlag för dialog om framtidens Stockholm – PM 2007-03-06. Stockholm City Council (Swedish).

Environmental and Real Estate Division, Stockholm City Council (2007). The Stockholm's Action Programme on Climate Change. Available at: http://www.stockholm.se/Extern/Templates/Page.aspx?id=172599 (last viewed on 1.8.2007).

Lidquist, A. (2005). Vilken soppa! En studie av processen bakom det politiska beslutet om av avvecklingen av Loudden och frågan om den framtida oljeförsörjningen i regionen. Department of Social and Economic Geography, Stockholm University (Swedish).

Malmsten, B. (2005). Förslag till hamnstrategi för Stockholm: Att hamna rätt. County Administrative Board of Stockholm (Swedish).

Swedish Environmental Protection Agency (1999). System med indikatorer för nationell uppföljning av miljökvalitetsmålen Rapport 5006. Swedish Environmental Protection Agency, Stockholm (Swedish).

– (2007). Swedish Environmental Goals. Available at: http://www.miljomal.nu/english/english.php (last viewed on 15.6.2007).

Transek (2005). Stockholmsregionens framtida oljeförsörjning: Etapp I – Inventering av nuvarande användning av oljeprodukter samt kartläggning av befintliga anläggningar och distributionssystem. Stockholm, December 2005 (Swedish).

– (2006). Stockholmsregionens framtida oljeförsörjning Etapp III – Slutrapport. Stockholm, December 2006 (Swedish).

The River Elbe in Hamburg

–

A description of the issues surrounding the dredging of the River Elbe and possible means of conflict resolution

Walter Leal[1], Agnieszka Hołda[1], Janne Juurikas[1], Irene Lucius[1], Dörte Krahn[1], Ahmed Quereshi[1]

Photo: *A. Holda*

Abstract

The development of the city of Hamburg has always been interlinked with and shaped by its harbour activities. Due to its inland location and topography, the port's main problem is the accumulation of sediments and the need for regular.

The further deepening of the River Elbe is also of on-going concern for the purpose of port development. An additional 1 metre depth of the River Elbe is vital to allow 24 hours accessibility of the port, independent of high tides, as well as to allow ships with 15.5 meters draught and bigger transport capacities than ever before to reach the port of Hamburg. This in practice means changes to the river bottom over a distance of 130 kilometres, from the port area downstream to the Elbe outlet into the North Sea. The further deepening of the river is envisaged for the years 2007-2009. For the Hamburg and its region this is a political decision of great importance creating controversies due to the often conflicting economic, environmental and social interests of different stakeholders.

---

1 TuTech Innovation GmbH – Hamburg – Germany.

This case study describes the background of the problem and provides different alternatives for its solution. A self assessment tool was introduced by the Hamburg team as part of this case study. This so-called 'Marker' was used for the assessment of the current state of Integrated Coastal Zone Management planning.

1. Introduction

From the historical perspective, Hamburg has always been bound to harbour activities, which have shaped the development of the city to a great extent. Due to its location inland and topography, the port has one problematic characteristic, the need for regular dredging of accumulating sediments.

The deepening of the river was specifically discussed in the 1990's, when it was discovered that the river sediments were highly contaminated, and needed to be treated before any disposal could be considered. The pollution was a result of upstream industrial activities in the former GDR and Czech Republic, which at that time largely disregarded environmental issues. Since then, many pollution prevention and control techniques have been implemented and the quality of the river water has improved. However, the polluted sediment is still tranported up and downstrean throughout the river.

Although some technical solutions to clean part of the dredged sediments have been found, further problems for the port of Hamburg have emerged. For example, the amount of sediment reaching the harbour had doubled by 2004, constituting over 8m tons, as compared with 4m tons in the years before. The capacity of on-land storage of sediments will be exhausted in 6 years, thus new sustainable long term solutions to deal with the increased amount of this material, including treatment and storage, need to be found. Moreover, recent observations followed by further research indicate a trend of decreasing water levels in the river as a result of climate change. Less water will affect both the people living in the Elbe river basin as well as the local fauna and flora, and additionally interfere with the port activities in the near future.

One of the on-going problems seen in the Hamburg area, which is highlighted by the issues mentioned above, concerns the further deepening of the Elbe River for the purpose of port development. To allow 24 hours accessibility of the port, independent of high tides, as well as to allow ships with 15.5 meters draught and bigger transport capacities than ever before to float to the port of Hamburg, the Elbe river needs to be deepened by another 1 meter. This in practice means changes to the river bottom over a distance of aprox.130 kilometres, from the port area downstream to the Elbe outlet into the North Sea.

Planned for the years 2007-2009, the further deepening of the river is a political decision of great importance for the Hamburg region, which creates controversies due to the often conflicting economical, environmental and social interests of different stakeholders.

This case study will therefore take a closer look at the multidimensional character of the Elbe River Deepening Plan by presenting different viewpoints of the stakeholders and their respective arguments for and against the plan. It will also discuss related environmental management issues which add up to the complexity of this case. Furthermore, a set of relevant conflict resolution methods, which bear in mind the circumstances that led to the current state of conflict, together with the presently drafted or prepared solutions (such as 'Concept for the sustainable development of the Tidal Elbe' as well as the German national ICZM strategy) will be described.

In addition, as part of the case study, the Hamburg team introduced a self assessment tool, called a 'Marker', which is used for the assessment of the current state of Integrated Coastal Zone Management planning. These developments will be outlined in this case study as well as lessons learned.

2. Problem background

2.1 Characteristics of the Elbe drainage area

The River Elbe is one of the largest rivers in Central Europe, third after the rivers Danube and Rhine in terms of length as well as catchment size. The drainage basin area of the River Elbe comprises 148,268 km^2 (IKSE 1995, 2000) and is shared between Germany, Czech Republic, Austria and Poland. However, Austria and Poland count less than 1% of the catchment area while 2/3 is located in Germany and 1/3 in Czech Republic. The source of the River Elbe is in the Giant Mountains („Krkonoše Mountains", Czech Republic), flows through the Czech Republic, through the northern and central part of Germany and discharges in the North Sea near Cuxhaven. The River Elbe covers a distance of 1091 km (727 km in Germany, 364 km in Czech Republic) (IKSE 1995, 2000) and along its way, the catchment area interacts with some of the major cities in the area, such as Prague, Dresden, Berlin and Hamburg (refer to Figure 1).

On its last 110 kilometres, before reaching the North Sea, the river is tidally influenced. Tidal fluctuations lead to very special forms of wetland habitat and biodiversity. This is also true around the city of Hamburg, especially where its container shipping port is situated.

Fig. 1: Characteristics of drainage area of the River Elbe

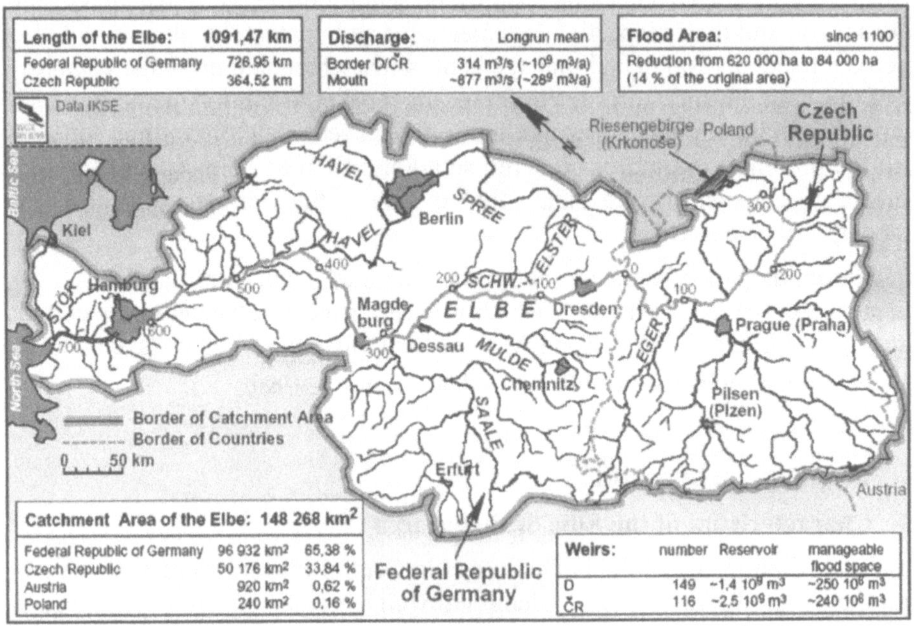

Source: *ARGE ELBE*

2.2 The city of Hamburg and the importance of its port

The Free and Hanseatic City of Hamburg is a city-state and the second largest city in Germany with 1.7 million inhabitants. It is a cultural and commercial centre for Northern Germany, its metropolitan region consisting of approximately 4 million people. Hamburg municipal area is equal to 755.3 km^2 whereas the metropolitan region covers a total surface of approximately 19.000 km^2 and embraces 14 districts around the City of Hamburg. After the Ruhr area and Berlin it is also the third biggest industrial area in Germany, with business related to: aircraft-, ship- building and automotive industry; electronics technology, precision engineering and optics industry; mechanical engineering; chemistry; mineral oil processing; and metal production. However, of highest importance to the city is the industry related to the harbour and the harbour activities per se (SAHSH 2005).

The Port of Hamburg is the largest port in Germany, the third largest port in Europe (after Rotterdam and Antwerp) and one of the ten largest container ports worldwide. It is considered a focal point for the trade conducted with Eastern and Northern Europe and is also an important connection point that ensures the traffic to and from the hinterland flows, thanks to its efficient network of inland waterways, feeder ships, railways and road carriers (HPA 2006a). The international trade volumes account approximately for one third of Europe's export. The surface area occupied for the activities relating to the harbour function constitute as much as 10% of Hamburg's total area.

The port and shipping sectors employ around 75,000 people in Hamburg. In addition, more than 133,000 jobs in a vast range of industrial and service sectors are directly or indirectly dependent on the port, which accounts for 12,7% of the total employment of the city generating a share of 14,4% to the GDP of Hamburg. In the metropolitan catchment area around 156,000 jobs depend directly or indirectly on the port and 258,000 throughout Germany (HPA 2007).

2.3 Accumulating sediment and the problem of previous contamination

In order to sustain the trade by allowing free passage of ships through the harbour, which is located around 110 kilometres from the North Sea, as well as to keep the harbour fully operational, the depth of the navigation route has to be assured. Like almost all harbours situated at tidal rivers, Hamburg harbour is affected by sedimentation problems. A frequent characteristic of harbour sedimentation is that 60-85% of the deposited material is located near the harbour entrance. This is mainly caused by flow-induced eddies due to the trapping effect of horizontal flow circulation. Sediments accumulate in the port of Hamburg through tidal action from the marine environment and through the deposition of fluvial sediments from the river Elbe and lead to a permanent decrease of water depth. Therefore, in the Port of Hamburg, dredging has to be undertaken regularly to secure a sufficient water depth for all types of vessel traffic.

Traditionally, the dredged sediments were used beneficially for land reclamation or agriculture. However, approx. 20 years ago the contamination of dredged sediments, mainly with heavy metals (As, Cd, Hg, Zn) and organic contaminants (PCB, Dioxins, PAHs), and the resulting negative effects on the environment, came into public focus (Heise 2005). Thus, it became clear that sediments needed to be treated before their usage or disposal in the Hamburg region. The pollution was a result of upstream industrial activities in the former GDR and Czech Republic (mining chemical, pharmaceutical, pulp and paper, as well as leather-processing industries), which at that time disregarded environmental issues (Netzband et al. 2002). This led to a broad political discussion. A Dredged Material Research Programme was initiated, but Hamburg could not act upstream to prevent further emissions into the river Elbe. The political change of the late 1980's led to significant improvements for the Elbe. Many industrial and agricultural complexes of the former communist regimes collapsed, and substantial amounts of discharges of pollutants ceased. The remaining industries and farms or those that had started since the early 1990's are generally equipped with modern pollution control technologies (Reincke et al. 2003). However, the polluted sediment is still tranported downstream by the river Elbe.

2.4 Solutions for the treatment of dredged material

Initially, a technical solution for the treatment of dredged sediment was devised, consisting of pre-treatment, which is the separation into sand and contaminated silt fraction, and the dewatering process followed by environmentally safe disposal of the silt in two specially constructed silt mounds (Netzband 2002). Pre-treatment is done in the large scale METHA plant (MEchanical Treatment and Dewatering of HArbour-sediments). It has a throughput capacity of 1 million m^3 of sediment per year. Its products, besides smaller amounts of coarse materials, are sand, fine sand and silt. The sand is used as construction material and is almost contaminant free (Detzner et al. 2004). The fine sand can be used in industry as a raw material or additive, whereas the silt can be used as a sealing material in the construction of disposal sites or can be used for harbour backfilling (Langaas et al. 2002). At present, a major portion of the dredged material generated by maintenance works is relocated into the Elbe. With the open water disposal, sediment coming from the river is put back into the aquatic system. The effects of relocation on the environment are minimized, in accordance with a concept of sustainable relocation.

The cumulative impact on Hamburg of both the amount of sediments from upper reaches of the Elbe and their contamination is that Hamburg has to spend substantial amounts of money on treatment, relocation and disposal of the dredged material. The costs for the city of Hamburg amount to roughly € 30 million per year, not including personal and capital costs (HPA 2005a).

2.5 Tidal pumping – more sediment reaching the harbour

Until the year 2000, the amount of sediment that had to be dredged every year was at a stable level of roughly 2 million m^3. Since then, the amount of material has been increasing, reaching 3 to 4 million m^3/year. In the year 2004, over 8 million m^3 of sediment needed to be dredged from the tidally influenced Hamburg area (HPA 2005b). The increase in the amount of sediment in the port is caused by a specific tidal pattern, the so called tidal pumping effect. It returns to the harbour the material that has been relocated in the past years in the area of the turbidity zone (HPA 2006c). In short, the high tide brings back more of the material to the port, as the low tide takes away. This is a problem of huge importance for Hamburg harbour, which not only has to dredge more accumulating sediments, but also has to find new ways to handle its treatment, as the current amounts exceed the capacities of the METHA sediment treatment plant. In addition, this also means an increase of the associated costs for the dredging itself and for treatment and storage of the sediment.

2.6 The problem of on-land storage space for treated sediments

Dredging activities and treatment of contaminated sediments are not the only problems with which Hamburg has to deal in view of increasing amounts of sediment reaching the harbour. It is also the issue of space that plays a crucial role here. Being a city state, Hamburg had to dispose of the treated sediments, originating from dredging, in its limited city borders. However, the on-land storage capacity for the treated sediments is predicted to satisfy the disposal needs only for the next 6 years (HPA 2006c). Therefore Hamburg had to seek out other solutions. The Hamburg Port Authority and the Waterway and Shipping Directorate („WSA Nord") for example, have developed a sediment management concept for the tidal Elbe, which was seen as an economically feasible solution for sediment storage. The plan encompasses storage of 4.5 millions m^3 of dredged material in the North Sea in the years 2005 to 2008.

2.7 Influence of climate change on Elbe water level

Observations of the fluctuations of the water level of the river Elbe indicate a decreasing water level trend. Further research carried out by the Institute for Climate Research in Potsdam as well as the Max-Planck Institute for Meteorology led to the conclusion that this is happening as a result of climate change, and moreover, show that the water level can decrease even further within the next years (DUH 2006; HA 2006). Less water can affect both the people living in the Elbe river basin as well as the port activities in the near future. With a decrease of water levels large vessels will no longer be able to reach the port of Hamburg. This will lead to severe consequences for the shipping trade and for the economy of the port as well as of the whole Hamburg region.

3. Conflict description

3.1 Reasons for deepening of the Elbe River

Accumulating sediment has always been a problematic issue for the harbour to deal with, and dredging the river for the purpose of providing good navigation can be described as a traditional activity. However, as shipping technologies change, the carrying capacity of vessels constantly increases, resulting in bigger draught of the ships. As a result, to sustain the trade and guarantee their competitiveness, harbours have to develop and adapt. Such changes, in the case of the port of Hamburg, also involve the deepening of the river Elbe.

In addition, at present, the navigability of the gateway from the Port of Hamburg towards the North Sea depends on the tidal fluctuations on the river Elbe. With reference to the Elbe, the low tide is longer than the high tide, 7 hours 30 minutes long. In Hamburg, the tidal range is 3.63m (2001–2005 average) (HPA 2007). If the tidal fluctuations are not taken in account, vessels reaching 12.80 meters fresh water draught may reach or leave the Port of Hamburg independently of tidal variations; while vessels with a deepest draught of 15.10m have to depend on the high tide. The outgoing vessels may use a 13.80m fresh water draught availability during the high tide to leave the Port of Hamburg (HHVW 1999). At present, the Lower Elbe is 15.30m deep, but for full navigability independent of the tide fluctuations, a deepening by a further 1 meter is necessary.

3.2 Economic, Environmental and Societal interests

As long ago as February 2002 the Senate presented Hamburg's application for a further adaptation of the Lower and Outer Elbe shipping channel to the Federal Ministry for Traffic, Construction and Housing (BWA 2004a). The development plan, whose start was initially set for the end of year 2007, has met a broad public debate and has faced criticisms from many non- governmental environmental organisations. The use, ecological state and further transformation of the river Elbe are important subjects for the different interest groups, which have various economic, environmental or societal aspects as their focus.

3.2.1 Economic interests

Some of the predominant economic interests are as follows:

- Harbour related activities are a central driving force for the economy of Hamburg and the whole of Northern Germany;
- Assuring competitiveness of the harbour on a European scale, with regard to trade as well as tourism;
- Utilising capacity and fulfilling the demands of the container terminals;
- Sustaining other smaller businesses, related to port activities;
- Investing in the development of the largest port in Germany.

Those interests are represented by the Hamburg Ministry of Economy and Labour Affairs, the Hamburg Ministry of Urban Development and the Environment, Hamburg Port Authority, the Hamburg Chamber of Commerce as well as tourism operators.

3.2.2 Environmental interests

Some of the predominant environmental interests are as follows:

- Protecting and maintaining the natural conditions of the Elbe environment in general;
- Keeping the morphological changes of the river on the lowest possible level, as they also impact on flooding;
- Lowering the industrial emissions of pollution from the harbour to zero;
- Acting against severe O_2 depletion, present during the summer time in Hamburg, which often results in a large number of fish kills;
- Ensuring that the disposal of potentially contaminated dredged sediment in the North Sea will not affect fish and the marine environment.

They are represented by Federal Ministry for the Environment, Nature Conservation, and Nuclear Safety, by various non-governmental environmental organisations such as Greenpeace, WWF, Deutsche Umwelthilfe, BUND, as well as by the general public.

3.2.3 Societal interests

Some of the predominant social interests are as follows:

- Securing employment for 146.000 people in the metropolitain region, directly or indirectly related to the harbour activities;
- Assuring the use of water for the purpose of social events, such as Elbe Swimming Day on 17 July 2005 (Elbebadetag 2005) or the Big Jump Event (Big Jump 2005) which now occur annually.

Social interests are represented by the general public as well as the Workers Union.

3.3 Divergent opinions on the deepening plans

Based on the different interests and stakes they represent, two conflicting groups advocating for or against the further deepening of the river Elbe can be distinguished.

The arguments for deepening of the Elbe are mainly supported by economic interests. Since the harbour activities are among the driving forces for the economy of Northern Germany, utilising their capacity is important and desired in the whole region. Also, further development of the port will not only promote environmentally friendly transportation of goods, but will also secure the current employment of 146.000 people, and related small businesses (BWA 2006). In addition, financial aspects speak for the deepening. As compared to other investments in the region, the costs of the plan, of around € 320 million, are seen as a relatively low investment. There is enough money in the national and state budget to carry out the plan, which is strongly supported by the Hamburg Ministry of Economy and Labour Affairs.

On the other hand, the arguments against the port development plan are various, and range from environmental concerns to suggestions for financing another deep sea water port.

The environmental arguments express the concerns of further morphological changes applied to the river and the consequences of the deepening to the oxygen content in the river, to fish populations as well as tidal fluctuations and sediment transport. The extent of the impacts can only be estimated, thus the real influence might prove to be more severe, as it was the case with the last Elbe deepening in 1999 (WWF 2005).

Based on recent research indicating that the general decrease of the water level in the Elbe is related to climate change, many non-governmental environmental organisations demand that a new feasibility study be carried out for the Elbe deepening plan, taking climate change aspects into account (DUH 2006). As the Institute for Climate Research in Potsdam predicts that low waters will be even more severe in the coming years, many NGOs claim that the 24 hours navigation availability to the harbour, independent of high tides, will not be achievable. Thus, in view of the increasing amount of sediment transported to the harbour, their suggestion to the local authorities is that the investment will not meet its purpose (LZ 2006).

The Federal Ministry for the Environment, Nature Conservation and Nuclear Safety does not deny that vessel transport is one of the most environmentally friendly ways of transporting goods. However, a major argument against the deepening is the long term vision of the container transport development. There are claims that after the year 2010, one deep water port will be developed on the German North Sea coast, serving as a hub for the current shipping industry (HA 2004a). Moreover, some argue that Germany is financially unable to develop more than one deep water port. Therefore, if the port of Wilhelmshaven receives the necessary financial support, it will become a serious competitor and a problem for the port of Hamburg in several years. Some experts confirm this prediction and speak for Wilhelmshaven: since it is situated directly at the coast, vessels would not have to travel at least 110 kilometres inland. Moreover, Wilhelmshaven does not have such extensive problems with sediments as Hamburg has.

In addition, an important argument used against deepening in 2007 is that despite predicted further increase in vessel draught and carrying capacity, further deepening actions of the Elbe will not be achievable (HA 2004b). The Hamburg tunnel under the river Elbe, for example, makes it technically nearly impossible (LZ 2004).

4. Relevant stakeholders and relevant conflict resolution methods

4.1 Categories of stakeholders

Based on the differing interests of the various parties concerned with the deepening plan, the coastal stakeholder community was identified and divided into three groups. The first category encompasses the relevant decision making bodies, whereas the second category focuses on interest based parties, whose fate is dependent on approval of the Plan. The third category gathers other institutions and research bodies.

Category 1: conflicting stakeholders

- Hamburg Ministry of Urban Development and the Environment
- Hamburg Ministry of Economy and Labour Affairs
- Federal Ministry for the Environment, Nature Conservation, and Nuclear Safety

Category 2: interest-based stakeholders
- Hamburg Port Authority
- Chamber of Commerce Hamburg
- Tourism operators
- Ship owners
- Port workers

Category 3: other stakeholders
- Universities
- NGOs
- Research Institutes
- Authorities dealing with integrated coastal zone management of neighboring Länder (German regions), in particular Schleswig-Holstein and Niedersachsen.

The interests of those stakeholders and their following viewpoints, for or against the Elbe Development Plan, have been described in the previous chapter. The next section deals therefore with the decision-making process and how that relates to conflict resolution.

4.2 Decision making process and conflict resolution

Having received the application for further adaptation of the „Lower and Outer Elbe shipping Channel" in 2002, the Federal Ministry for Traffic, Construction and Housing set up a project group for carrying out the required preliminary investigations. A set of feasibility studies was seen as a precondition for the project to be put on the top priority list of the Federal Traffic Routes Plan. They included investigations of the technical feasibility of another shipping channel expansion, an environmental risk study, as well as an economic cost-benefit analysis. The results showed that another expansion is technically feasible, economically cost-effective and ecologically justifiable (BWA 2004b). On this basis in September 2004 the Federal Government gave its preliminary consent to the Development Plan, which left the path clear for the preparation of the Plan Approval Procedure. In specific terms it meant that (BWA 2004a):
- detailed plans for the project were to be completed;
- measures leading to the award of a contract for the environmental studies had to be carried out;
- scoping meetings were to be held;
- parameters of the environmental sustainability study were to be determined;
- involvement or participation of those authorities and associations that are responsible or affected was to be assured;
- and environmental studies as such were to be carried out.

The environmental studies consisted of in-depth hydrological studies, carried out by the Federal Institute of Hydraulic Engineering. In addition, the landscape conservation plan was initiated to define the compensating ecological measures that are to be called for, as well as an intensive 'transparent planning' strategy to keep the general public well informed. It is thus assumed that an independent survey as specified by the cabinet resolution (BWA 2004a) will give full consideration to the natural conservation aspects. The above described studies were to be brought to a conclusion in time for the Plan Approval Procedure to be initiated in the middle of the year 2006.

With regard to such big development plans as the Hamburg Port development, the disputes of conflicting stakeholders are highly political. Therefore, the decision making process, which includes the Feasibility Study and Plan Approval Procedure, takes into consideration different perceptions and concerns of the interested parties as well as ensuring public participation, which together should lead to the resolution of the conflict.

In addition to the legislative process, various research institutes and universities carry out both independent in-depth studies on a given subject, as well as providing additional and often new input and long term predictions of the factors that are relevant to the case. An example can be the two already mentioned institutes – the Max-Planck Institute for Meteorology in Hamburg and the Institute for Climate Research in Potsdam – making studies on climate change and its influence on the Elbe River. Providing reliable data on a highly disputed subject leads to transparency and contributes to conflict resolution, whereas new information and observations assure that all aspects or impacts in the Development Plan are being considered.

5. Possible solutions or lessons learned

There is no single, satisfactory solution to such multidimensional problems, where conflicting interests and different stakeholders are involved. The outcome of the decision making process is unlikely to satisfy all parties to the same extent, unless a sustainable long-term vision for the region as well as for the port is taken into account. That is why as one of the means for conflict prevention and management, through better development of the coastal region, a Sustainable Development Plan for the tidal Elbe taking into account the national Integrated Coastal Zone Management Strategy, was developed.

Moreover, in addition to the legislative Plan Approval Procedure, several other decision support tools can be used for conflict management and resolution. The Marker self-assessment tool and the SWOT Analysis, as well as the DPSIR Framework, which were applied to this case study by the COASTMAN Hamburg team, can be used to enhance the process by improving its transparency and by analysing it from a different perspective while at the same time making sure that all interests are taken into consideration.

5.1 DPSIR Framework

As part of the Hamburg case study an overview of environmental management dimensions of current and potential future conflicts in the river Elbe in Hamburg was developed using a DPSIR Framework. In addition, conclusions on how to improve the environmental condition of the river Elbe in Hamburg, taking the social, economic and political development of the region into consideration, were drawn. Applying this framework provides a higher transparency to the assessment of the origins and consequences of environmental problems, thus serving as a good conflict management tool.

5.1.1 Description of methodology

Adopted by the European Environment Agency, the DPSIR framework is a crucial instrument for describing the interactions between society and the environment and relationships between the sources and results of environmental problems. DPSIR stands for:

- *Driving Forces*, the human or economic activities that cause pressure – e.g. population increase, increased urbanisation;
- *Pressures*, expression of the driving forces – e.g. emissions of harmful substances;
- *State of the environment*, which describes the quality and quantity of natural resources – e.g. nitrogen and phosphorus levels in inland and coastal waters;
- *Impacts*, effects or loss of use experienced in the environment – e.g. loss of amenity in a waterway, decrease in biodiversity;
- *Responses*, various corrective actions undertaken that may affect any of the inputs of the conceptual framework – e.g. EU Directives, taxes, incentives.

Relationships between the components are described in Figure 2. It describes the interaction between the State of the environment, the anthropogenic Pressures and the underlying direct and indirect Driving Forces. Further, the Impacts of the changes in the State of environment and the Responses from society in order to counteract these unwanted impacts are comprised by the framework. The DPSIR model illustrates both that the societal activities affect the state of the environment, and that there is a feedback from the State of the environment to society in the form of environmental policy initiatives for individual sectors (agriculture, transport, industry etc.). In order to understand their dynamics it is also useful to focus on the links between the DPSIR elements. For instance, the relationship between Driving Forces and Pressures by economic activities is a function of the eco-efficiency of the technology and related systems in use. Similarly, the relationship between the Impacts on humans or eco-systems and the State depends on the carrying capacities and thresholds for these systems. Whether society „Responds" to impacts depends on how these impacts are perceived and evaluated and the results of Responses on the Driving Forces depends on effectiveness of the Response (Arevalo & Heise 2003).

Fig. 2: DPSIR assessment framework

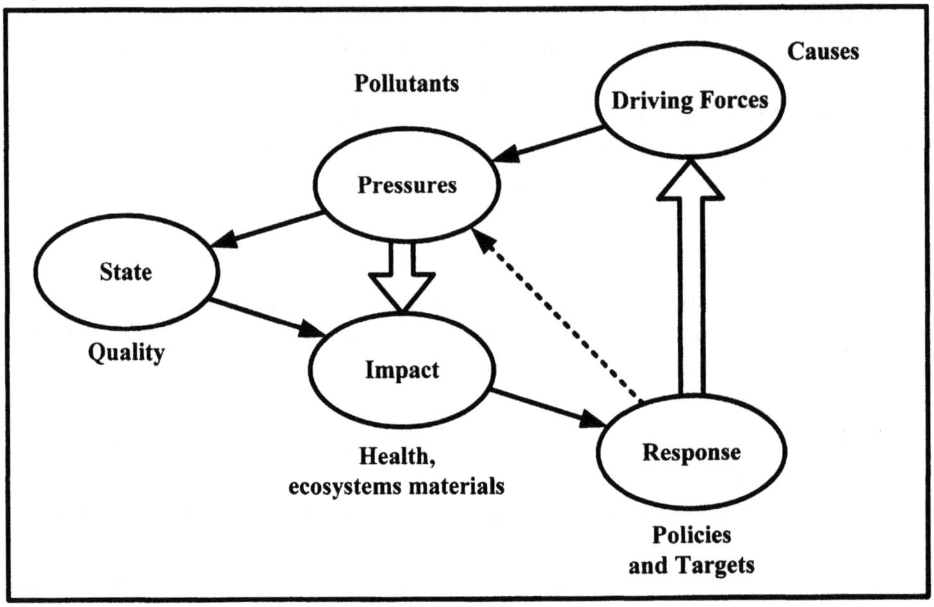

Source: EC, 2003

5.1.2 Application of the DPSIR Framework to the River Elbe

General overview of the Driving Force – Pressure – State – Impact – Response framework for the Hamburg part of the river Elbe was carried out outlining different aspects of the environmental problems and conflict. Results are presented in Table 1.

Tab: 1. DPSIR Framework applied to River Elbe in Hamburg.

Driving Force	Pressure	State of environment	Impact	Response
Port Activities and Navigation	– Dredging – Utilisation of the port area – Shipping accidents	– Change of river morphology and ecological condition of the river – Pollution (oil, chemicals)	– Disappearance of several species in flora and fauna – Contaminant accumulation in sediments and also in fishes – Treatment and disposal of dredged material made difficult due to the heavy contamination of sediments with heavy metals (especially Cadmium)	– National, European and international policies – National and international cooperation agreements and conventions
Flooding	– Flood protection	– Change of river morphology and ecological condition of the river		
Households and Industries	– Point and diffused pollution sources	– Pollution of the river water and sediments with nutrients and hazardous substances		

5.1.3 Future environmental management problems and related conflicts of the River Elbe

Based on the assessment of the current problems and trends, possible future conflicts can be identified. These may arise from several environmental problems, which have to be faced and taken into consideration while managing the river Elbe. They are as follows:

- A continuing threat to biodiversity is the future decrease and degradation of the habitats as a result of climate change, further specialisation of agriculture and continued increase of traffic infrastructure and industrialisation.
- The proliferation of new alien species, mainly as a result of new conditions caused by increasing water temperature, ballast water and introduction by man may pose future problems.
- Future water quality problems with new hazardous substances may arise. Endocrine disruptors are substances that have disruptive effects on hormonal processes in organisms including man. PCBs and TBT are examples of endocrine disruptors, the production and use of which are being or have been phased out or are strictly limited.
- Due to the development in the shipping sector (e.g. insufficiently trained crews, sub-standard vessels, and inadequate salvage capacity) there is a potentially increasing risk of shipping accidents.

- Another future threat is a potentially accelerating sea level rise, which may on the one hand have an impact on the ecosystem, and on the other hand pose an increasing burden on coastal protection and the safety of the hinterland.

The solution of future conflicts depends on the extent to which the barriers to progress in environmental protection and sustainability can be overcome. This is mainly due to the complex, inter-sectoral, inter-disciplinary and international nature of both the problems and the solutions. These barriers are underpinned by shortcomings in institutional structures, non-implementation of commitments already made and lack of information on and understanding of possible 'win-win' solutions for achieving sustainable outcomes. Such solutions embrace competitiveness and innovation, social cohesion, territorial cohesion and the protection and maintenance of scarce natural resources and valuable ecosystems (EEA 2004).

5.1.4 Conclusions drawn from application of DPSIR Framework

The application of the DPSIR Framework to the case study of the River Elbe provided higher transparency of the studied problem and has led to one conclusion: a sustainable river management strategy of the river has to consider integration and implementation of sectoral strategies with regard to policies/activities in the coastal zone and mainland. As there is still insufficient or lack of cooperation between local, regional, national and EU authorities in the preparation, implementation, enforcement and coordination of the rules and regulations, cooperation between responsible authorities is essential. In order to facilitate sustainable management the rules and regulations should be harmonised and simplified. While acknowledging that public participation in policy and management already exists on many levels it is encouraged to further enhance bottom-up processes wherever possible and as far as compatible with the principles of parliamentary democracy.

The full report from the application of the DPSIR Framework to Hamburg case study can be viewed on COASTMAN website at: www.coastalmanagement.net

5.2 SWOT Analysis

Within the framework of the COASTMAN project a conflict assessment method, the SWOT analysis (**S**trengths, **W**eaknesses, **O**pportunities and **T**hreats) of the Port of Hamburg has been performed. It casts a different light on the positive trends and challenges in the environmental management of the Elbe in Hamburg.

5.2.1 Application of the SWOT Analysis to Hamburg case study

Table 1 depicts the strengths, weaknesses, opportunities and threats of the environmental management components of the river Elbe. The commitments to abide by national laws and the EU Water Framework Directive (WFD) as well as the geo-economic importance of the port are positive signs for the future of

Hamburg's harbour areas. These strengths make it possible to ensure the ecological sustainability of the river Elbe and the economic stability of the Port of Hamburg. The location of sources of pollutants outside the boundaries of Hamburg and the non-availability of an economical dredged sediment disposal option are serious weaknesses found within the framework of the analysis, which may also be a source of conflicts. At the same time, the globalisation of the Eastern and Central European countries has provided an opportunity for the Port of Hamburg to promote its business.

In contrast, the proposed harbour development plans have been opposed by different stakeholders on the basis of environmental, ecological and social reasons. The lack of harmonisation of ideas on the development plan poses a major threat to the Port of Hamburg.

Tab: 2. SWOT Analysis of environmental management components of River Elbe. Own illustration

Strengths	Weaknesses
The concerned authorities and stakeholders are eager to find solutions to the technical, environmental and legal problems being faced by Hamburg's coastal areas. The commitment to national laws and international conventions is very strong among the authorities responsible for coastal resources management. Hamburg has the ability to ensure its economic stability, due to its well-established port economy. Hamburg has efficient hinterland links. Hamburg has been the European centre for trade with East Asia and China. Hamburg has very close relations with the Baltic Region and Eastern Europe.	Major sources of contaminants do not lie within Hamburg's jurisdiction. From upstream, different pollutants travel and stick to sediments, resulting in contaminated sediments at a later stage. Deepening of harbour project is questioned due to the presence of contaminants in Dredged Material (DM). The Federal Government does not support Hamburg with regards to its idea of DM management. The disposal of DM is a very expensive task. Hamburg has to spend a large amount of money on the treatment of DM and its disposal. The lack of land available for disposal of environmentally safe dredged material disposal and relocation restrictions are also weaknesses.

Opportunities	Threats
An opportunity exists to improve the environmental status of coastal water by implementing EU directives at European level.	Lack of consensus between Hamburg and other upstream Federal States regarding the management of the Elbe catchment area is a major threat.
The EU encourages research work in the field of coastal resources management. It provides a great understanding of specific problems.	Furthermore, NGOs are also opposed to the deepening of the harbour.
Hamburg has the potential to compete with other ports in the region.	The relocation of dredged material and changes to the river morphology pose a threat to the ecosystem.
Hamburg has the opportunity to benefit economically from the economic development in Eastern European countries and Russia.	Emissions from old polluted sites in coastal water will lead to the ongoing pollution of surface water and sediments.

5.2.2 Conclusions from applying the SWOT Analysis

The analysis has disclosed the various strong and weak points related to the activities of the Port of Hamburg. Within its framework, different aspects of the environmental management of the harbour were investigated to establish its capacity to face present and future economic, environmental, and social challenges. Keeping in mind the different operational aspects of the Port of Hamburg, the following recommendations have been put forward to avoid or at least minimise possible conflicts among different stakeholders:

- Long-term solutions to reduce contaminants in the sediments from upstream must be developed with the help of the Federal Government and other Federal States. This will help to ensure the deposition of clean sediments from the Port of Hamburg in suitable places.
- More extensive environmental impact assessment of the deepening of the Port of Hamburg may be carried out in order to determine its precise impacts on the biodiversity of the Elbe.
- A long-term solution to environmentally friendly sediment disposal should be drawn up to alleviate the fears of different stakeholders.
- The possibility of sub-aquatic disposal should be investigated.
- Legal obligations can be met by following a long-term policy as mentioned above. In addition, close co-operation with other stakeholders will reduce the chances of any legal conflicts arising.
- A close coordination with the NGOs and other concerned bodies is necessary to keep alive the social and cultural value of the Elbe.

5.3 German National ICZM Strategy

In the view of the European Union, the implementation of an ecologically sustainable, economically balanced and socially compatible coastal zone management that also gives careful consideration to cultural aspects meriting protection and maintains the integrity of the coastal ecosystems is of decisive importance for sustainable development. By virtue of the recommendation of the European Parliament and the Council of 30 May 2002 (EC 2002) to implement a strategy for integrated management of coastal zones in Europe, the Member States were called upon to focus their attention on this tool and submit a report by February 2006.

The development of a national ICZM strategy in Germany is based on an assessment of the economic, social and ecological situation of Germany's coastal zones as well as of the legal, political and administrative structures and institutions that have an influence on the conditional framework for taking action in the coastal regions.

For the national strategy ICZM is viewed as an informal approach aimed at supporting sustainable development of the coastal zones through good integration, coordination, communication and participation. On the one hand, ICZM is a process that should permeate all planning and decision making levels as a guiding principle, and on the other hand, it is a tool applied prior to formal procedures for the purpose of integrated identification of potential development and conflict as well as for resolving conflicts. The German national strategy is based on the following basic principles:

- ICZM shall promote *sustainable development* of coastal zones with their specified ecological, economic and social features and support the sustainability strategy of the Federal German Government;
- ICZM represents a guiding principle for political and social action at all levels in coastal zones and is aimed at coordinating the development of coastal zones through a comprehensive approach and *integration* of all concerns;
- ICZM incorporates all relevant policy areas, economic and scientific stakeholders, social groups and levels of administrations into the process (*participation*) in order to identify development potential at an early stage, find solutions for which there is a consensus and improve conflict management;
- ICZM is viewed as a continuous process that combines the phases of planning, implementation and evaluation of changes in coastal zones so as to make the best possible use of experience for the future (*experience transfer*).

The results of the assessment of strengths and weaknesses of the current situation in coastal zones indicate that major aspects of the ICZM basic principles have already been implemented in important areas through the planning practice established in Germany, on respective legal foundation. However, it is also plain that further steps have to be taken to improve the achievement of goals at the planning and in particular at the individual decision making level. With respect to the existing set of tools and activities the national strategy envisages four areas in which further steps should be pursued:

- Further optimisation of the set of legal instruments according to the basic ICZM principles,
- Creation of the basis for continuation of the dialogue process,
- Best practice projects and their evaluation,
- Development and application of ICZM indicators.

However, as the German national strategy states, it should not lead to the creation of new bureaucratic obstacles. On the contrary, by initiating broad-based participation, acceleration of the problem-solving process is targeted since conflicts can be identified, discussed and resolved at an early stage.

Implementation of the national strategy will have to be pursued by Federal and state legislators as a top down approach, especially in connection with optimisation of the existing set of tools in accordance with the basic ICZM principles. Furthermore, the Federal and Länder governments will have to provide resources and communication platforms and perform coordination tasks. In addition, another focal point will be the bottom up oriented cooperation between governmental institutions with local, regional and social actors.

The Integrated Coastal Zone Management Strategy for Germany has been presented by the Cabinet on 22 March 2006, and was officially submitted to the European Commission at the beginning of April (Umwelt 2006) (BMU 2006).

5.4 ICZM and the Marker self-assessment tool

One of the keys to the sustainable development of the City of Hamburg is reconciling the interests of those stakeholders developing its harbour with the interests of other stakeholders present, such as tourism, fisheries, housing, and nature conservation among others. Therefore within the Hamburg case study an investigation was carried out on how instruments of Integrated Coastal Zone Management can contribute to resolving conflicts.

5.4.1 Aims and objectives

The reasons for applying the ICZM Marker in the framework of the Hamburg Case Study were the following:

- The Marker was introduced to COASTMAN partners and to representatives of the Hamburg stakeholder community and was regarded as an interesting instrument.
- A contribution to the European discussion process on ICZM instruments was sought by testing the Marker in Hamburg and providing feedback to the European level and the Federal Ministry for the Environment, Nature Conservation and Nuclear Safety.

- Application of the Marker in form of a workshop facilitates the discussion process among coastal stakeholders about what ICZM is, what has been done in the recent past in order to establish it as part of the local, regional or national coastal management and planning process, and where the weaknesses are, thereby contributing to conflict resolution and communication among the stakeholder community.

On 28 April 2006, the National ICZM strategy for Germany was presented at a conference in Bremen. The strategy itself as well as several conference participants was critical about the European ICZM Progress Marker. This is why it was decided to focus less on the results of a Marker application in Hamburg and more on achieving consensus on how the Marker could be improved in order to best serve its purpose of indicating ICZM progress in Germany and facilitating conflict management in the course of a discussion process, active stakeholder participation, and integration of different interest groups.

5.4.2 Description and methodology

The method of choice for bringing together the stakeholder community and initiating a discussion process on conflicts and achievements of coastal resource use was the application of the European indicator for measuring progress in Integrated Coastal Zone Management (ICZM), an element of implementing the ICZM Recommendation by the European Parliament and the Council of 2002 (EC 2002).

The „Progress Indicator", often referred to as „Marker" because of its simple, binary response structure, was developed under the guidance of the EU ICZM Expert Group and its Working Group on Indicators and Data, established by DG Environment of the European Commission. The Working Group received assistance from the Topic Centre for Terrestrial Environment, a structure of the European Environment Agency, and EUCC – The Coastal Union as an additional subcontracted consultant. The resulting Progress Indicator/Marker has received approval of the EUC ICZM Expert Group and was recommended for testing in EU Member States on various levels (Expert Group ICZM 2004).

The ICZM Progress Marker is an instrument for assessing to which degree ICZM in its widest sense is being implemented on local, regional and national level and whether there is progress over time (e.g. within five-year intervals) in applying and establishing the ICZM approach. It highlights success stories as well as failures, thereby helping to focus resources where they are most needed.

A second set of indicators has also been developed by the Working Group on Indicators and Data under the guidance of the Expert Group. This Indicator has the complementary aim of assessing sustainable development along the coast, thereby showing whether progress in ICZM has the intended results (Working Group ID 2004).

The ICZM Progress Marker methodology (included as Annex 1 to the full report) consists of 31 actions characterising the process from an early stage of ICZM (phase 1: Aspects of coastal planning and management are taking place) to the final stage (phase 4: An efficient, adaptive and integrative process is embedded at all levels of governance and is delivering greater sustainable use of the coast). This list of actions is complemented by annotations that describe in more detail what each action implies. The goal is to assess whether this action has already been completed or not or whether information is lacking and therefore an assessment is not possible. This assessment is to be carried out for the local, regional, and national level and applied for the year 2000 and the year 2005 in order to indicate progress over time. The results can be visualised in form of a table where positive assessments are marked with green, negative ones with red, and „don't know" with yellow (please see Annex 1 contained in the full report).

A representative group of coastal stakeholders should make the assessment in the framework of a workshop. Optionally, the assessment can be done in written form to familiarise the stakeholders with the methodology and provide an indication as to which actions are being assessed the same way by all stakeholders and for which ones a joint discussion is necessary in order to reach a common conclusion. In a second step, a workshop should follow.

Application of the Marker, in tests carried out so far, has shown that civil servants working in central government departments, for example, will not necessarily have much knowledge of what is going on locally. Similarly, local practitioners will have restricted knowledge about what, if anything is happening at regional or national levels. Even people working in the same organisation often differ with their colleagues in assessing whether a particular action is being fully implemented or not.

Therefore, it is beneficial to bring together practitioners from different administrations, organisations, agencies and interest groups to jointly complete the table. In this way, a more accurate picture of how far ICZM is being implemented at all three spatial levels – national, regional and local can be gained.

In fact, the act of completing the table is itself an important step in helping to implement ICZM. The debate necessary to decide on an answer, even one as apparently simple as 'yes' or 'no', leads to an exchange of opinions about which organisations and agencies are doing what on the coast, and to what effect.

5.4.3 Application of the Marker in Hamburg

In April 2006, identified stakeholders received a letter inviting them to do the self-assessment by filling in the ICZM Progress Marker table translated into German and consulting with the Explanatory Notes, where necessary. They were also informed that they would be invited to a workshop in Hamburg in order to discuss conflicting issues and achieve a consensus on how to assess progress of

ICZM in the city. After several weeks, those stakeholders that had not responded were contacted by phone and e-mail in order to discuss reasons for not participating and answering questions.

On 26 June 2006, a workshop took place at TuTech Innovation in Hamburg with the participation of 20 stakeholders. Several others had voiced interest but could not participate due to other obligations. The scope of the workshop had been extended to include the issue of Sustainability Indicators in the context of the German ICZM Strategy, because of its complementary nature to the ICZM Progress Indicator.

Irene Lucius from EUCC – The Coastal Union presented the specific objectives of the workshop in detail and briefed the participants on the latest ICZM policy development on European level. She stressed that the ICZM Progress Marker has been applied in the context of national and regional workshops in Belgium, France, South UK, Poland, Spain, Malta, and Italy. The results so far are promising: There is a clear positive trend over the past 5 years and the indicator has been well received, although some measures for improving the tool have been put forward, e.g. the need to add „in progress" to the already existing possible categories of „Yes", „No", and „Don´t know" for assessing the 31 actions.

In addition, Bastian Schuchardt, bioconsult, gave a briefing on the state of the German ICZM process, referring to the German ICZM Strategy, which recommends, among others, the establishment of an ICZM Secretariat and ICZM Forum. The Federal Ministry for the Environment, Nature Conservation and Nuclear Safety had contracted bioconsult to develop proposals for coastal sustainability indicators tailored to the German situation.

Sustainability Indicators had been developed as a follow up to the European ICZM Recommendation and applied within the framework of the project DEDUCE in some countries. Achim Daschkeit presented results of applying those indicators to German regions and visualising them with the help of WebGIS. The problems encountered in the process were in particular the low availability of data for some indicators and the lack of clear objectives, limits and consequently judgment for interpreting the results (e.g. what is a „sustainable" or „unsustainable" population density along the coast?). These norms have to be defined by the coastal stakeholder community first. Also, the direct relevance of some of the indicators to ICZM has not been made clear. He added that the application of the EU sustainability indicators is not sufficiently relevant from a policy point of view.

Achim Daschkeit continued by providing his analysis of the ICZM Progress Indicator and the process of applying it. The development of ICZM processes – planning or regional development processes – can be evaluated with progress (also called process) indicators. In Achim Daschkeit´s opinion, the EU ICZM process indicator is not very appropriate for the German context because it does not deliver precise information about the ICZM process and because the „by-product" of doing the assessment – raising awareness on what ICZM is and how it is being implemented – is not important in a country where stakeholders are already well informed about the approach. Many actions – in particular those of the first phase – could best be assessed by an expert doing literature search for legal acts and policy initiatives. The other actions require a subjective assessment for which standard empirical methodology exists in the form of opinion surveys with a range of possible answers (e.g. from „I fully agree" to „not at all"). Intensity and quality of communication and cooperation, for example, are best assessed with this methodology if the usual conditions and limitation of empirical social research are being taken into consideration.

Another weakness according to Achim Daschkeit is the fact that the composition and minimum number of stakeholders applying the marker for a particular area are not defined. Consequently, results are not well comparable and only of limited value. Furthermore, the formulation of some questions is too vague. He concluded by saying that the content of the ICZM Progress Marker is appropriate, but the process of application needs to be refined.

5.4.4 Results of the ICZM Progress Marker test run in Hamburg

The ICZM Progress Marker table was sent to 55 stakeholders from 45 institutions representing different sectors and administrative levels. Most of them are working for Hamburg institutions, some representing the „Länder" Schleswig-Holstein and Niedersachsen and at Federal Government level. Out of those, only four did the assessment, while some explained why they could not do so in written form. The most commonly cited reasons were unclear responsibility, lack of time, the negative attitude of the recently published ICZM Strategy for Germany towards the European ICZM Marker, and the special status of Hamburg (it is both a municipality and a „Land" and it is located not directly on the coast).

While the number of respondents is too low for a proper statistical analysis, some qualitative remarks can be made: There was considerable agreement in the positive judgment of actions in the first phase and least agreement on the assessment of ICZM progress in Hamburg from the year 2000 to 2005.

5.4.5 Discussion

The Workshop participants agreed on the following weaknesses of the Marker:

1. Many aspects of ICZM are already established in Germany, but they are not called ICZM. Therefore, the Marker tends to undervalue progress in those actions where the term ICZM is mentioned.
2. The Marker is too extensive and not clear enough in some points. Therefore its application can lead to long discussions about terminology without pro-

viding much insight. It would be better to limit the Marker to a maximum of 10 crucial points.

3. Neither the actions themselves nor the annotations are formulated clearly enough. Some actions consist of different elements that need to be assessed independently. This leads to inaccuracy, redundancies, and frustration during the application process.
4. Some actions can be assessed only subjectively (e.g. there is no objective answer to what is „properly staffed and properly funded")
5. "Yes", „No", and „Don't know" are not specific enough for assessing the actions.
6. The aims and target groups of the Maker should be better communicated. It should be made clear for example that it is not an instrument for solving local problems.
7. The terminology must be better defined. Some terms such as „stocktake of the coast" or „report on the State of the Coast" are clear in the English language context but not yet established in the German language, neither as English nor as German terms.
8. Filling in the Marker table individually does not lead to meaningful results because of ambiguous terminology and because left to themselves, stakeholders tend to answer questions purely from the perspective of their interests and discipline.
9. Clear instructions are needed which type of stakeholders and how many of them must contribute to the assessment. The group of people who ended up applying the marker in Hamburg appears too random.
10. The ICZM Sustainability and Progress Indicators are being discussed and applied independently at the moment although they are (or should be) interrelated.

The participants recommended the following improvements:

1. The conditions for applying the marker need to be defined clearly. Two alternative models have been proposed:
   a) If an ICZM Forum is established with representation of the major stakeholder groups (as the Germany ICZM Strategy recommends for the near future), the Marker can be applied during a Forum session. This would provide the opportunity to define terminology and, during the discussion process, achieving a common opinion on the state of ICZM, integrating the perspective of the different sectors and disciplines.
   b) If no Forum (or comparable body) is established, the Marker can be split into two groups of actions: those that can be assessed objectively with the help of a literature search, and those that can be assessed with the

help of a questionnaire answered by a sufficiently large and representative group of stakeholders in written form.
2. Sustainability and Progress Indicators should not be assessed independently.
3. The actions need to be formulated more clearly and if necessary, terminology needs to be better defined without using the term ICZM. Thereby, the annotation can be reduced to a large degree or made obsolete.
4. Some actions need to be split into two or more elements, to be assessed independently. Other actions, of lesser relevance to ICZM, should be eliminated or integrated into others in order to reduce redundancies and reduce overall number of actions.
5. More communication efforts are needed to convince stakeholders about the necessity of ICZM. This would increase motivation for applying the Marker.
6. The aims of applying the Marker have to be made very clear.

Some of the participants voiced their willingness to engage in further development of the Marker on the European and German level.

5.4.6 Conclusions for the ICZM progress Marker test run in Hamburg

The ICZM progress Marker test run in Hamburg has shown that there are many objections to the methodology in Hamburg and the region. It would not be advisable to let individual representatives of the stakeholder community carry out a written assessment as in its present form. Also, in the German translation, the wording of the actions is not precise enough. A workshop type setting would be beneficial with sufficient time allocated – probably one day as minimum – to explain to the participants the precise purpose of the exercise, to clarify terminology, and to moderate a discussion process between the different stakeholders that leads to a common assessment.

When interpreting these results it is important to bear in mind that this was a test run of the application of the ICZM marker. The purpose of this workshop seemed to be unclear at the beginning and led to some confusion.

While in some European regions starting the process of stakeholder interaction is a goal in itself, the participants in Hamburg seemed much more goal-oriented. This is why more emphasis would need to be placed on explaining the reason for the testing exercise. Once the goal – constructive criticism in order to improve the Marker – was made clear, the German audience focused entirely on this aspect, not on testing the methodology and interpreting preliminary results.

Superficially, it looks like the COASTMAN objective of applying the Marker as a mechanism for conflict resolution by facilitating a productive discussion process among stakeholders about achievements and challenges of the ICZM process has failed. At a second glance, however, the process helped to unite stakeholders in their sometimes strong criticism of the methodology and to raise their interest in contributing to developing ICZM methodology for Germany. During the workshop criticism arose e.g. regarding the formulation and grouping of the marker actions, leading to an intense discussion, which left some participants irritated at times.

Ideally, the planned German ICZM forum will become a platform for developing and applying the Marker in the near future. If this effort fails, application of the Marker will require a strong champion – such as the Federal Ministry for the Environment, Nature Conservation and Nuclear Safety – who could bring stakeholders together for Assessment Workshops.

The discussion process has also brought into the open that the stakeholders, who came together in Hamburg, have their own, not very clear definition of ICZM. This may also be the case in other countries, and would mean that in some cases considerable adjustments to the way coastal resource use is being managed and planned would be required. The fact that the tedious, meticulous and often long-lasting planning process with stakeholders participation so common in Germany is a good base for, but not the equivalent of ICZM, has to be communicated much better. In any case, as long as ICZM is an ambiguous and vague term in any country's context, the application of the Marker will have its problems.

Independent of these aspects, the discussion process has revealed several shortcomings of the present methodology and it can be hoped that critique and recommendations will be taken into consideration during future rounds of revising the assessment tool.

In particular, the idea of producing a clear protocol for applying the Marker should be considered. If ICZM progress assessments are to be made comparable between different European countries, regions and municipalities, the assessment process should be agreed upon. Developing two options – the joint assessment in the framework of a workshop and the written assessment, split into a literature search part and a questionnaire – may be considered as a pragmatic but at the same time sound solution. A critical revision of the actions, in particular defining key terminology more clearly and reducing inaccuracies and redundancies, would be recommended.

5.5 Sustainable development plan for the tidal Elbe

In June 2006, the Hamburg Port Authority in cooperation with the Waterway and Shipping Directorate WSD Nord) prepared a „Concept for a sustainable development of the Tidal Elbe River as an artery of the metropolitan region Hamburg and beyond" (HPA 2006c). This policy paper discusses concepts, which will provide overall holistic and topical inputs to the preparation of an Action Plan for Sustainable Development of the Tidal Elbe. It shows that there are real chances of developing a plan for the tidal Elbe, from which benefits for different interests, such as fisheries, agriculture, nature protection, tourism, shipping and economy will arise.

The document describes the specifications and the situation of the tidal Elbe. It explains future strategies and their general frameworks, together with an analysis of the political framework and the current need for investigation and action. It is made clear that the involved Länder (German states) and the Federation have to jointly take the necessary long-term measures, in order to positively influence the dynamics of the river system and its morphological and ecological features.

The successive implementation of methods over the next 100 years is expected to lead to the development of a dynamic estuary, in which various ecological and economical functions will be improved. The identified core aspects of the future sustainable Action Plan for the tidal Elbe are the following:

- Attenuation of the increasing energy of the tides through technical measures focused on the estuary funnel;
- Creation of tidal zones in the area between Glückstadt and Geesthacht;
- Optimisation of sediment management, taking into consideration the whole Elbe river system.

The concept recognises that without adequate measures, the system of the tidal Elbe will increase its land area, creating not only ecological disadvantages but also making the maintenance of the water bodies and Hamburg harbour more complex. In addition to the planning of measures, different models will be used to simulate processes of the river system, which are expected to lead to convergence of development of the tidal Elbe with nature protection.

The sustainable development element is seen as a challenging, demanding and long term task. This project, which encompasses the next hundred years, can only be successful if all interests are taken into consideration. In this view, the authorities submitted the developed concept for discussion, comments and public inspection.

6. Conclusions

The development plan for the Lower and Outer Elbe, which envisages the deepening of the river by a further one meter, has met a broad public debate and has generated criticisms from various non- governmental environmental organisations. The use, ecological state and further transformation of the river Elbe are important subjects for different interest groups having related economic, environmental or societal aspects at their focus. The conflicting interests have generated virile political pressures, and a dilemma can be seen between the positive, but relatively short, economical consequences, and the long term, sustainable planning for the region, where investing in development of another deep sea port is highly likely.

Concerns were expressed about further modifications of the river environment when some of the long term aspects influencing the decision making were not taken into consideration, as well as when the national port development plan is not being considered. Therefore, in addition to the legislative decision making process, which includes the Feasibility Study and Plan Approval Procedure, a Sustainable Development Plan for the tidal Elbe as well as a German national Integrated Coastal Zone Management Strategy, were recently developed.

However, to ensure consideration of all different perceptions and concerns of the interested parties several other decision support tools can be used, which will in turn lead to better conflict management and potentially towards conflict resolution. Presented in this report, the Marker self-assessment tool, the SWOT Analysis, as well as the DPSIR Framework, which were applied to this case study, are recommended to be used to enhance the decision making process by improving its transparency, while at the same time making sure that all interests are taken into consideration.

It is hoped that the presented methods will be useful to other harbour cities and might lead to cooperation and an exchange of experience in conflict prevention and conflict management.

7. References

Arevalo, E. & Heise, S. (2003). Sustainable Solutions to the Sediment Management on the River Basin Scale: The Elbe River Case under specific consideration of risk management and communication aspects. 2nd SedNet-Conference, September 2003.

ARGE ELBE (2006). Das Einzugsgebiet der Elbe. Available at: www.arge-elbe.de/wge/Einzug/EinStart.html (last viewed on 14.08.2006).

Big Jump (2005). Big Jump for Living Rivers. Available at: www.rivernet.org/bigjump/ (last viewed on 14.08.2006).

BMU (2006). Integrated Coastal Zone Management in Germany. Assessment and steps towards a national ICZM strategy. Bundesministerium für Umwelt, Naturschutz und Reaktorsicherheit. March, 2006.

BWA (2004a). Extension of the Lower and Outer Elbe in Hamburg. State Ministry for Economic and Labour Affairs. Press release, 16.09.2004. Available at: www.hafen-hamburg.de/en/index2.php?option=content&task=view&id=141&Itemid=130 (last viewed on 14.08.2006).

– (2004b). Fairway adjustment: Hamburg has done its homework! New Elbe Deepening is Technically Feasible, Economically Cost-effective, and Ecologically Justifiable. State Ministry for Economic and Labour Affairs. Press release, 23.02.2004. Available at: www.hafen-hamburg.de/en/ index2. php?-option=content&task=view&id=170&Itemid=130 (last viewed on 14.8.2006).

– (2006). Fahrrinnenanpassung der Elbe auf gutem Wege. State Ministry for Economic and Labour Affairs. Press release from 02.06.2006. Available at: http://fhh.hamburg.de/stadt/Aktuell/pressemeldungen/2006/juni/01/2006-06-01-bwa-fahrrinnenanpassung.html (last viewed on 14.08.2006).

Detzner, H.-D., Netzband, A. & Knies, R. (2004). Dredged Material Management in Hamburg. Terra et Aqua 96 (September), pp. 3-14.

DUH (2006). Klima setzt Elbschifffahrt auf Grund.Umweltverbände fordern Neubewertung der Elbeasbau- und Unterhaltungspläne. Deutsche Umwelthilfe. Press release from 16.01.2006. Available at: www.duh.de/pressemitteilung.html?&tx_ttnews[pS]=1150373389&tx_ttnews[pointer]=14&tx_ttnews[tt_news]=452&tx_ttnews[backPid]=457&cHash=c458f45456 (last viewed on 14.08.2006).

EC (2002). Recommendation of the European Parliament and of the Council of 30 May 2002 concerning the implementation of Integrated Coastal Zone Management in Europe (2002/413/EC). Full text on http://eur-lex.europa.eu/pri/en/oj/dat/2002/l_148/l_14820020606en 00240027.pdf.

– (2003). Reporting for Water – Concept Document: Towards a Shared Water Information System for Europe (WISE). Common Implementation Strategy for the Water Framework Directive. November 2003.

EEA (2004). EEA Signals 2004. European Environmental Agency. Copenhagen.

Elbebadetag (2005). Elbe Swimming Day, 2005. Available at: www.elbebadetag.de (last viewed on 14.08.2006).

Expert Group ICZM (2004). Minutes, Expert Group ICZM Recommendation, 4th Meeting, 24.11.2004. Full text on http://ec.europa.eu/environment/iczm/pdf/minutes_24_11_04.pdf.

HA (2004a). Elbvertiefung: Trittins Nein empört Hamburg. Hamburger Abendblatt. Press release from 11-12.09.2004, p. 1.

– (2004b). Elbvertiefung: Nur Punktstieg für Hamburg. Hamburger Abendblatt. Press release from 16.09.2004, p. 1.

– (2006). Elbe wird weniger Wasser haben. Hamburger Abendblatt. Press release from 17.01.2006, p. 27.

HABAK-WSV (1999). Handlungsanweisung für den Umgang mit Baggergut im Küstenbereich (HABAK-WSV). Available at: www.bafg.de/servlet/is/ 5830/ HABAK_WSV_1999.pdf (last viewed on 14.08.2006).

Heise, S. (2005). Studie zur Schadstoffbelastung der Sedimente im Elbeeinzugsgebiet. Ursachen und Trends. Hamburg, Hamburg Port Authority, p. 195.

Hesse, M. (2002). Logistics real estate markets: indicators of structural change, linking land use and freight transport. Paper for the ERSA 2002-Conference: From Industry to advanced Services. Dortmund, Free University of Berlin, Department of Geography. May 2002. Available at: www.ersa.org/ersa-confs/ersa02/cd-rom/papers/094.pdf (last viewed on 14.08.2006).

HHVW (1999). Port of Hamburg Handbook 1999/2000. Hafen Hamburg Verkaufsförderung und Werbung AG (May 1999). United Kingdom, Compass Publication Limited.

HPA (2005a). Baggergutmanagement, Hamburg Port Authority, 2005.

– (2005b): Situation der Baggergutverbringung in Hamburg und Sediment-bewirtschaftungskonzept an der Tideelbe – Aktueller Handlungsbedarf. Bericht der Abteilung: Prozess Wassertiefen, Hamburg Port Authority, Freie und Hansestadt Hamburg 01.04.2005.

– (2006a). Port Development Plan. Efficient hinterland links via all carriers. Hamburg Port Authority. Available at: www.hamburg-port-authority.de/index.php?option=com_content&task=view &id=152&Itemid= 139&lang= english (last viewed on 14.08.2006).

– (2006b). The importance of the port for the Hamburg metropolitan region. Available at: www.hamburg-port-authority.de/index.php?option= com_content&task=view &id=149&Itemid=135&lang=english (last viewed on 14.08.2006).

– (2006c). Concept for a sustainable development of the Tidal Elbe River as an artery of the metropolitan region Hamburg and beyond. Available at: www.hamburg-port-authority.de/images/stories/download/Strategiepapier_ Tideelbe_komplett_english_final_051006.pdf (last viewed on 10.11.2006).

– (2007). The Port of Hamburg – Facts and Figures 2007. Available at: www.hafen-hamburg.de.

IKSE (1995). Die Elbe und ihr Einzugsgebiet. Internationale Kommission zum Schutz der Elbe. Dresden, IKSE.

– (2000). Die Elbe von 1990 bis 2000 – 10 Jahre erfolgreiche Zusammenarbeit in der IKSE. Internationale Kommission zum Schutz der Elbe. Dresden, IKSE.

Langaas, S., Aliakseyeva, N., Gooch, G., Lopman, E., Nilsson, S., Timmerman, J. (2002). Working Paper on Environmental Information in Transboundary River Basin Policy-Making and Management: Selected European Case Studies. October, 2002. Available at: www.lwr.kth.se/publikationer/ PDF_Files/MANTRA_East_WP.pdf (last viewed on 14.08.2006).

LZ (2004). Ein Tritt vor, ein Trittin zurück. Scharfe Kritik an neuer Hürde für Elbvertiefung. Lüneburger Landeszeitung (LZ). Press release from 17.09.2004.

– (2006). Warnung vor Elbvertiefung. Lüneburger Landeszeitung (LZ). Press release from 13.01.2006, p. 15.

Netzband, A. (2002). Part 2: Treatment and Confined Disposal of Dredged Material. Dutch-German Exchange (DGE) on Dredged Material.

Netzband, A., Reincke, H. & Bergemann, M. (2002). The River Elbe – A Case Study for the Ecological and Economical Chain of Sediments. Journal of Soils & Sediments 2(3), pp. 112-116. Available at: www.htg-baggergut.de/ Downloads/JSS-3-112-116.pdf (last viewed on 14.08.2006).

Reincke, H., Hurst, S. & Schneider, P. (2003). Strategy Concept Elbe. Available at: www.piewak.de/pdf/13-elbe.pdf (last viewed on 14.08.2006).

SAHSH (2005). Monitor Wachsende Stadt. Bericht 2005. Statistisches Amt für Hamburg und Schleswig-Holstein. Available at: www.statistik-nord.de/ fileadmin/download/allgemeines/Bericht_Monitor_ Wachsende_Stadt.pdf (last viewed on 14.08.2006).

Umwelt (2006). Integriertes Küstenzonenmanagement. Chance für unsere Küsten und Dialogplattform für die Zukunft. Bundesministerium für Umwelt, Naturschutz und Reaktorsicherheit. Umwelt, no. 6-2006, p. 357.

Working Group ID (2004). Measuring Sustainable Development on the Coast, a Report to the EU ICZM Expert Group by the Working Group on Indicators and Data led by ETC-TE, Report 4/2004. Full text on http://ec.europa.eu/environment/iczm/pdf/report_dev_coast.pdf, Annex: http://ec.europa.eu/environment/iczm/pdf/report_dev_coast_an.pdf.

WWF (2005). Die Elbvertiefung von 1999. Tatsächliche und prognostizierte Auswirkungen. WWF Deutschland 2005. Available at: www.wwf.de/imperia/md/content/pdf/fluesseundauen/4.pdf (last viewed on 14.08.2006).

Spatial Conflict Resolution and Coastal Zone Management for Haapsalu Bay

Arvo Iital[1], Krista Vilta[2], Enn Loigu[1], Jüri Kurba[2]

Photo: Tõnis Padu

Abstract

Haapsalu Bay is part of the Baltic Sea and extends deeply into the land in the western part of Estonia. The inner part of the bay is characterised by intensive eutrophication, growth of reed and other macrophytes, decreasing open surface water area and sedimentation. Environmental objectives for Haapsalu Bay have not yet been set up. However, different water users (e.g. recreation, industry, fishery sector, nature conservation) require a water quality in accordance with the quality standards and norms for their respective purposes. Reduced possibilities to satisfy all water users of the bay with their specific needs regarding the water quality and quantity can cause additional stress and conflicts. Therefore, if the aim is to maintain and improve the ecological status of the bay, not only do environmental measures have to be implemented, but also more complex solutions will need to be developed.

1 Tallinn University of Technology, Department of Environmental Engineering, Tallinn, Estonia.

2 Haapsalu Town Government.

This case study set out to facilitate spatial panning in the area, develop a costal zone management strategy and measures for protection of the bay and its ecological potential. In addition, in order to resolve potential conflicts arising from the use of environmental services a forum was initiated between different stakeholders using scenario methods, discussions and roundtables to create citizen participation.

1. Introduction

Haapsalu Bay is part of the Baltic Sea and deeply penetrates the western part of Estonia (Fig. 1). The inner part of the bay is a typical brackish water body characterised by intensive eutrophication, growth of reed and other macrophytes, a decreasing open surface water area and sedimentation that has lead to raising the seabed in this part of the bay. Therefore, maintenance of shipping routes requires continuous dredging of the channels. This case study aimed at assessing the main conflicts between different water users and other stakeholders with variety of economical, ecological or social interests. One of the tasks involved evaluating the investment in waste water technology in Haapsalu town from a system view. Before 1998 there was a traditional waste water treatment plant (WWTP) in Haapsalu town, the biggest settlement in the bay catchment, with a very low treatment efficiency. Only mechanical treatment of sewage waters was carried out in the plant, which served about 30% of the town population. The reconstruction of the WWTP was finalised in 1998 when a traditional chemical treatment stage to remove phosphorus was implemented. In 2001 additional nitrogen removal technology was introduced. Even though treatment efficiency is much higher now it is still not sufficient due to highly concentrated sewage water in the inlet of the WWTP, which makes it difficult to achieve necessary quality levels in the outlet. This was caused by a remarkable decrease in per capita water consumption by the population. The latter is mainly due to the increased price for potable drinking water after investments in the waste water handling technology.

The agricultural pollution load decreased considerably over the past 15 years due to reduced amounts of fertilisers applied and decreased numbers of livestock in Estonia. Despite these changes and the overall decrease in pollution load to the bay the ecological quality of the bay has not been improved and the bay is subject to continuous environmental pressure. Massive eutrophication in the bay, characterised mainly by enlarged reed-bed areas, is still the most acute problem. Environmental objectives for Haapsalu Bay have not yet been set up.

However, different water users (e.g. recreation, industry, fishery sector, nature conservation) require a water quality in accordance with the quality standards and norms for their respective purposes. Reduced possibilities to satisfy all water users of the bay with their specific needs regarding the water quality and quantity can cause additional stress and conflicts. Therefore, if the aim is to maintain and improve the ecological status of the bay, not only do environmental measures have to be implemented, but also more complex solutions will need to be developed. This also means that during the planning process potential social effects of development plans and programmes have to be considered.

Fig. 1: Map showing the location of the case study area

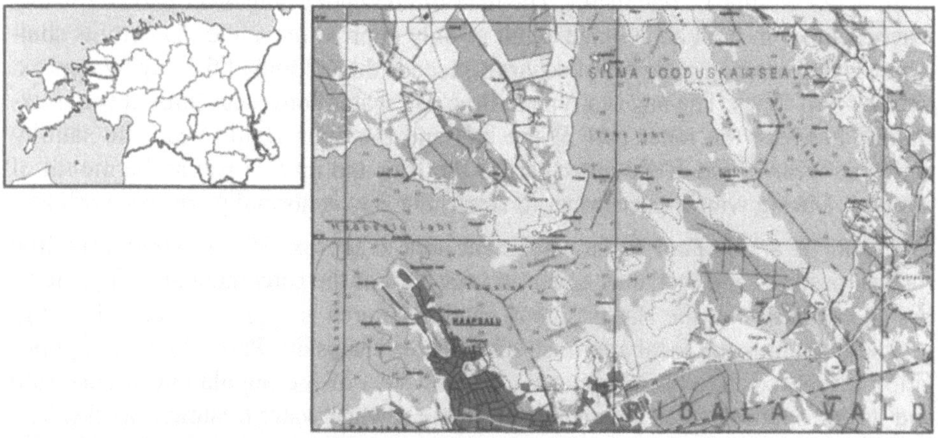

The more specific tasks of the work included:

- Facilitating spatial panning and developing proposals for a costal zone management strategy and measures of protection for the bay and its ecological potential taking into account that currently the applicable water quality standards for designated uses are being failed to be maintained.
- Demonstrating ways of spatial conflict resolution and initiating a forum between different stakeholders in order to resolve potential conflicts in the future, and to provide opportunities for public participation in all aspects of coastal zone management.
- Identifying critical coastal areas and land uses in the catchment, which, individually or cumulatively, may contribute significantly to a degradation of coastal waters.

Reaching these goals required the assessment of the pollution load to the bay, the ranking of the importance of different pollution sources, and the development of tools and proposals for improvement of the ecological quality of the bay. The latter included using scenario methods, methods for developing indicators for sustainable development, discussions and roundtables to create citizen participation for the future.

2. Study area and background of the problem

The catchment area of the bay is relatively flat with absolute height reaching only 15-20 meters above sea level. The surface area of the bay is 47 km^2 and the total catchment area is 412 km^2. The inner part of the bay is very shallow with a maximum depth of approx. one meter. The maximum depth of the outer part of the bay reaches 4 to 5 m with an average depth of only approx. one meter. Due to its shallowness the mean annual river discharge exceeds the volume of the bay by approx. two and half times. The bay is a typical brackish water body with limited water exchange between the inner part of the bay and the open Baltic Sea. The salinity varies between 1 psu in the inner part of the Bay and up to 7 psu in the mouth of the Bay. During winter (December to April) the Bay is normally covered with ice.

The share of arable land in the catchment area is approx. 25% (10,000 ha) while forests, pastures and wetlands cover around 72% of the catchment area. The number of inhabitants in the Haapsalu bay catchment area is approx. 17,000. The biggest settlement is the local administrative center Haapsalu (Photo 1), with a population of approx. 12,000 inhabitants. Around 80% of the population in Haapsalu and its surrounding areas are connected to the sewage water treatment works. The WWTP in Haapsalu was thoroughly renovated during the late 1990s and early 2000s. In 2001 the nitrogen removal technology was introduced.

Photo 1: View of Haapsalu town and the bay

Photo by Tõnis Padu

The town is a well-known recreational area and seaside resort in the peninsula. The curative mud used in local sanatoria is widely spread in the seabed sediments. The amount of curative mud in the two main bay areas Tagalaht (38.45 ha) and Voosi (6.2 ha) accounts for 165,000 and 15,000 m^3, respectively. Annually approx. 400 tons of mud are excavated. The demand for mud is increasing. The growth of reed (*Phragmites australis*) and other macrophytes in the bay is intensive (Photo 2).

Photo 2: Increasing reed-bed area in the bay

Photo by Arvo Tarmula

Reed belts cover approx. 1,400 ha (nearly 30%) of the Haapsalu Bay area and provide a remarkable storage for nitrogen and phosphorus. The bay is an important area for migrating birds as well as spawning fish (Photo 3). The inner part of the bay is part of the Silma Nature reserve area.

Photo 3: The bay is an important area for migrating birds as well as spawning fish

Photo by Arvo Tarmula

Agriculture in the catchment area of Haapsalu Bay is at a relatively low intensity. The main industry in the area is food production. A large number of people work in tourism and the health care sector, e.g. in local sanatoria, as well as in the public sector, e.g. schools etc. The main point pollution sources of the Bay are municipal waste waters and sewage waters from the food industry.

3. Main stakeholders and relevant conflict resolution methods

The list of the main stakeholders in the study area includes:

- *County administration* responsible for planning on county level.
- *Coastal municipalities* responsible for planning on local level.
- *Environmental authorities* responsible for good state of coastal environment including management measures.
- *Developers* responsible for carrying out of environmental impact assessment if necessary.
- *Fishery organisations* responsible for sustainable fishery planning.
- *Farmers* responsible for cost-effective agricultural management, sustainable land-use and use of resources.

3.1 County administration, coastal municipalities and environmental authorities

Coastal municipalities and environmental authorities are responsible for the maintenance of a high quality of life and high quality coastal environment, the development of the area and for the implementation of relevant management measures. This includes delivery of high quality drinking water to all users, improved waste water treatment efficiency and a decreased point source pollution load to the bay. The treatment of natural water to guarantee the quality necessary for potable water required additional investments. Consequently, these investments raised the price for the water and caused conflict between the goal of reducing overall water consumption of the general public and industry, and fulfilling national and HELCOM regulations regarding waste water treatment efficiency and overall impact to the water quality of the bay. The factors determining the water pricing policy are ecological, economical and social. According to the EU Water Framework Directive all ground, surface and coastal water bodies should be of at least Good ecological quality by 2015. Fulfilment of this requirement may also require additional resources. At the same time the pricing must take into account the ability of water consumers to pay for the water, because all people have a right to consume high quality drinking water. Thus, both the water works as well as the local municipalities should bear in mind these potentially conflicting interests of different parties and stakeholders. Therefore, public participation is essential in achieving necessary environmental as well as social targets. Coastal municipalities and environmental authorities have to find answers to questions such as:

- How will different scenarios of infrastructure investment affect future water use?
- What will be the impact of different water pricing structures, as well as financial, economic, social, cultural and environmental factors on future water use?

Despite biological treatment introduced in the late 1990s and an additional nitrogen removal stage in 2001 the treatment efficiency of the Haapsalu WWTP (Photo 4) remained quite stable being only 18.8% in 2004 comparing to 24.2% in 1995 and 17.3% in 1996 (Iital et al., manuscript). The mean concentration of total nitrogen (TN) is still high at 56.4 mg N/l in the incoming waster water stream and 45.8 mg N/l in the effluent in 2004. The annual load of total phosphorus (TP) decreased by more than half, from approx. 4.1 tons in 1995/96 to 1.9 tons in 2003/04 (Iital et al., manuscript). The decrease in annual load of BOD_7 has also been remarkable. In 1995/96 the annual load from the WWTP was 57 tons, while in 2004 it was only 4.6 tons.

Photo 4: Aeration tank (left) and secondary settlement tank (right) of the Haapsalu Waste Water Treatment Plant

Photo by Haapsalu Water Works

The WWTP was dimensioned for a mean discharge of 7,220m^3/d. The real discharge has been much lower, with an average of only approx 2,300m^3/d in 2004. The amount of sewage water delivered to the WWTP dropped by 30% from 1,159 thousand m^3 in 1988 to the level of 843 thousand m^3 in 1996, and remaining on that level being 857 thousand m^3 in 2004. This decreasing trend was accompanied by continuous drop in the overall per capita consumption of potable water. The per capita use of water decreased from 175 l/d in 1994 to 70 l/d in 2004 which has resulted in increased concentrations of nutrients in both the wastewaters delivered to WWTP and especially in nitrogen concentrations in the effluent. Having the mean influent concentration of nitrogen on the level of 50-60 mgN/l, like in

Haapsalu, it is quite unrealitic to achieve the target values for effluent concentrations (e.g. 15 mgN/l) using conventional technologies for nitrogen removal with an average treatment efficiency of 50-60%. After all, the concentration of TN still exceeds the level of 15 mgN/l. The treatment efficiency for TP is quite close to the required level, but still not sufficient, especially regarding maximum allowed effluent concentrations that was 2.1 mg P/l in 2004. The reason for high effluent concentrations is partly in very high TP concentrations in the influent and probably also in insufficient treatment technology. Having a goal to keep the price for the water for the population as low as possible the chemical precipitation is not used in the summer period in the Haapsalu WWTP.

The annual riverine nutrient load to the bay accounted 172 and 193 tons of nitrogen and 8.9 and 3.6 tons of phosphorus in 1995/96 and 2003/04, respectively (Iital et al., manuscript). Remarkable part of, both TN and TP, in the rivers origin from agricultural diffuse sources, e.g. 43% and 58% of TN (74 and 123 tons) and 57% and 39% of TP (5.1 and 1.5 tons), respectively in 1995/96 and 2003/04 (Iital et al., manuscript).

Tab. 1: The annual riverine nutrient load to the bay

	1995/96, tonnes y^{-1}		**2003/04, tonnes y^{-1}**	
Non-point nutrient source	**TN**	**TP**	**TN**	**TP**
Diffuse agricultural load	69.7	3.2	123	1.5
Diffuse load from natural areas	96	3.3	67	1.5
Point source load	2.5	0.5	2.8	0.6
Total	**168.2**	**7**	**192.8**	**3.6**

Indicators and actions necessary for resolving conflicts:

3.1.1 Investments to improve environmental quality

A source apportionment of pollution loads indicates critical investment needs for Haapsalu municipality and rural municipalities. More precise calculations should be made in the strategic planning process.

3.1.2 Effectiveness of investments in WWTP purification technology

A cost-effectiveness analysis of investments in purification technology showed that additional efforts should be made especially to reduce the phosphorus load on the bay. This is likely to improve the nitrogen and phosphorus ratio in the bay. The results have to be implemented in the strategic planning process. All chemical and biological processes that enhance the potential self-purification capacity in water bodies including shallow bays need further studies in the future. There will probably be a need to use the self-purification capacity of water bodies through natural or artificial wetlands like reed-bed areas of the Haapsalu Bay.

3.1.3 *Improved capacity of WWTP*

The WWTP must implement the new permitted emission values for outlet water from the plant according to the regulations for treatment facilities serving more than 10,000 people equivalents. The data will be used in the planning process.

3.1.4 *Integration of environmental aspects with planning process and solving conflicts*

Spatial planning is carried out at four levels in Estonia: national level, county level, municipality level and detailed planning on local level. The purpose of strategic planning (national master plans, county plans, general plans of municipalities and towns) is to balance territorial, economic, social and environmental as well as individual and general interests. The purpose of detailed planning is to prepare building and development activities while considering the potential environmental impacts. While influencing public opinion as well as training of administrative officers, specialists, interest groups and the public, e.g. shaping public preferences, special attention is paid to the introduction of transparency and democratic requirements.

The national strategy plan „Eesti 2010" was adopted in 1995 in Estonia. All counties and cities have also adopted their strategic development plans. Municipalities must approve general plans by 2007. Additionally, Noarootsi municipality in the northern part of the Haapsalu Bay catchment area is preparing a supplementary coastal area plan. The work on theme planning „Environmental objectives in spatial planning and land-use" on county level is ongoing. The theme plans have two sub-themes: „Green Network" and „Valuable Landscapes". To date, detailed planning has been carried out only in Haapsalu town. In other municipalities (except Haapsalu) of the task area the work has been delayed partly due to the lack of skilled personnel (e.g. architects). The lack of detailed plans results in uncontrolled changes in land-use, especially in coastal areas used for summer resorts. According to the Estonian Environmental Strategy 2010 one of the goals is to achieve and maintain good quality of the coastal sea for the biota as well as for human needs. The required actions include decrease in the amount of pollutants discharged to the costal sea and sufficient treatment of sewage waters. Implementation of environmental management systems by the public sector is another target set by the Estonian Environmental Strategy. Further development of environmental indicators to evaluate the results set up by the environmental strategy is still needed. The strategic document also points out that further clarification is needed as regards the environmental impact assessment of plans and programmes.

Haapsalu town has adopted a development plan for 2004-2009, revised in 2004 and in 2006, that includes among others:

- Strengths and weaknesses (e.g. lack of sustainable development plan – Haapsalu agenda, lack of environmental overview) in different environments.
- The list of activities and needs for changes.

- An action plan that will be updated depending on the needs and funding.

One of the main objectives for the revised plan is the development of recreation in coastal areas as well as in the vicinity of other water bodies. The idea is to make coastal areas more attractive for the people including a specific action plan covering a period of 5 years and ideas on how the coastal zone should look like to meet recreational needs.

There are several problems in the planning process in Estonia that cause a conflict between various interested parties:

- When implementing detailed plans strategic plans are not always followed;
- Balanced protection of public interests and environmental aspects is not always granted;
- Environmental targets are not set in many cases;
- The quality of plans is varied;
- Public participation in the planning process is often insufficient. According to the 2003 Planning Act all interested parties and people must be involved in the planning process (§ 4 and § 16).

The technical assistance to local governments and the public will hopefully help to implement the necessary additional management measures and to maintain or improve the ecological quality of the bay.

3.2 Environmental authorities (Silma Nature Reserve)

Silma Nature Reserve (covering 4,780 ha) was established in 1998. The area includes the inner part of Haapsalu Bay east and north-east from Haapsalu town and the coastal areas. The nature reserve is an important nesting area for water birds and a spawning area for freshwater fish. Intensive eutrophication has led to an increased reed area. Efforts to curb this development have created an increased push for using reed as a construction material or for incineration. The cleaning of the bay from reed also is expected to improve nesting conditions for birds and spawning conditions for fish but may result in a negative impact on the open bay if the capacity of the bay to serve as an effective filter, sink, and transformer of inorganic nutrients and eroded material is reduced.

These sometimes controversial interests of reed users, fishermen, local population and recreation need careful environmental impact assessment.

Indicators and actions necessary for resolving conflicts:

3.2.1 Changes in the number of nesting water birds

The number of nesting water birds depends on the preservation of suitable conditions for the birds. It includes eutrophication control of the bay and suitable solutions regarding the use of reed. Eutrophication in the inner part of the bay is mainly caused by changes in agricultural activities. Therefore, it is vital to find solutions that will satisfy both agricultural producers in maintaining sustainable agriculture in the catchment area of the bay, as well as the environmental authorities who are responsible for preserving suitable conditions for the biota in the nature reserve area.

3.2.2 Changes in fish stock

Results from the study regarding the significance of different pollution sources and possible pollution load scenarios for the future should be used by local authorities and fishery organisations to manage fish resources in the bay and in rivers.

3.3 Fishermen

Fishermen use the bay area for fishing and as a transportation route. The use of ships requires periodic dredging of the shipping channel in the bay due to the shallowness of the bay and intensive sedimentation. Sedimentation is caused by eutrophication and solid material transported to the bay via rivers. The dredging will cause conflicts between ecological/recreational interests and fishing/yachting interests of the local people and visitors of health resorts. Fishermen are interested in maintaining a proper income from the fish catch that depends on the conditions in the bay. The water quality can deteriorate due to municipal discharges of sewage waters and diffuse load of nutrients via rivers. Therefore, being interested in maintaining the high water quality fishermen need to know what will change in the state of the bay due to changes in land use, economy, pressure by tourism, etc., and what the impact of these changes might be on the future fish stock.

The increase in fishing and yachting activities requires proper design and operation of the marina hull maintenance areas. It is a significant way to prevent the entry of toxic pollutants into ground and marine waters. Therefore, specific impervious areas (e.g. cement areas) for hull maintenance activities such as painting should be maintained in the harbours. The harbours must be capable of collecting wastewater from ships. In the surrounding areas the share of impervious surfaces should be kept as low as possible to minimise storm water pollution directly to the sea. The increased impervious surface area, compaction of soils and canalisation of storm water result in an increased volume and peak discharge of runoff (Hood and Iital, 2006). The higher volume/peak flow rate of runoff leads to an increased export of suspended solids and associated nutrients and metals.

Indicators and actions necessary for resolving conflicts:

3.3.1 Changes in fish stock

Results from the study regarding the significance of different pollution sources and possible pollution load scenarios for the future should be used by local authorities and fishery organisations to manage fish resources in the bay and in rivers. Hobby fishing in the Silma Nature Reserve is regulated by the Regulation of the Minister of Environment No. 75 of 15.12.2005. Fishing is not allowed from April 1 to September 19 and is restricted by a licensing system in other seasons. Since Haapsalu Bay is an important fish spawning area, it would be reasonable to use the water quality standards of Regulation 78/659/EEC for its fishing waters. At present, the water quality of the bay does not comply with this regulation.

3.3.2 Changes in composition and abundance of bottom dwelling biota

Results from the study regarding changes in the abundance and composition of bottom dwelling fauna should be used to evaluate the potential of having high-quality fish in the coastal zone.

3.4 Farmers

Farmers are usually interested in improving agricultural management practices to decrease their possible impact on the environment, including the bay. The problem is whether the bay can act as a huge bio-pond in the future taking care of „agricultural" nutrients and providing suitable conditions for other water users (fishermen, tourists, nature conservation, and recreation). Enlargement of the reed bed area will increase pressure to use it as construction material or for incineration. Therefore, the ecological potential of higher vegetation to act as a trap for nutrients in the bay should be carefully assessed before its removal from the bay.

According to studies (Brandt et al. 2003, Iital et al, manuscript) the riverine load of nitrogen on Haapsalu Bay even increased from the mid-1990s. This increase was caused mainly by intensified agricultural production after the depression in the 1990s and by the increased share of arable land actually in use. The use of inorganic nitrogen fertilisers for agricultural crops in Lääne county increased more than sixfold from 1996 to 2004, and the area specific fertilisation rate with inorganic fertilisers almost doubled from 78 kg/ha in 1996 to 122 kg/ha in 2004 (Statistikaamet 1997; Statistikaamet 2005). A similar increase took place as regards the use of organic fertilisers, from 24 tons/ha in 1996 to 35 tons/ha in 2004 (*Ibid.*). The fertilised area increased fivefold within the same time period. Assuming that those changes in Lääne county match those in the Haapsalu Bay catchment area, the intensification of agricultural production is evident. The significant decrease in total phosphorus riverine load can be explained by the implementation of agricultural-environmental measures, especially in farming practices and manure handling that may indicate an increased environmental concern of farmers.

It is highly likely that after the area of cultivated land is expanded and agricultural production grows, accompanied by a higher rate of fertiliser application, the diffuse nitrogen load from agriculture will increase. Assuming that the increased nutrient load does not cause immediate problems in the bay due to its high buffering capacity, more efforts should be made to implement the best agricultural practices accompanied by other legal, organisational and agro-technical measures. These actions will minimise the possible increase of agricultural load and the risk of failure in achieving the targets set up by the WFD.

Indicators and actions necessary for resolving conflicts:

3.4.1 Sustainability of agriculture in the area

With the knowledge that there is a clear link between coastal water quality and land use activities along the shore, the sustainability of the agricultural sector was assessed using data about inputs to the sector and nutrient balances on farm level for selected farms and for a selected small agricultural catchment (Rägina). The inputs and outputs of nitrogen and phosphorus on farm level as well as on catchment level are now more or less balanced. This means that there are much better conditions for getting higher yields and better income for farmers. This slightly positive balance should be maintained in the future and the rapid increase of fertilisation rates should be suppressed to keep the overall nitrogen and phosphorus load to the bay at the present level. Proposals for a more cost-effective management and regulations regarding sustainable use of resources (e.g. fertilisers) were introduced during stakeholder forums.

3.4.2 Introduction of environmental management systems in agriculture

The introduction of environmental management systems in agriculture is a target set by the Estonian Environmental Strategy. The idea of and possibilities for the implementation of environmental management systems were introduced to some agricultural producers. Possible environmental indicators for evaluating the efficiency of measures to achieve the results set up by the environmental strategy were introduced during the seminars.

3.5 Coastal population

The coastal population in the area can be divided into two groups: population of Haapsalu town and rural population. Inhabitants of Haapsalu town are more or less affected by increased tourism to the area, and by planning and developing processes. Their living conditions depend also on the quality of Haapsalu Bay, treatment efficiency of the WWTP and the water pricing policy. The rural population is interested in preserving the beautiful agricultural landscapes and environmentally friendly agricultural production. The income of the rural population is very much determined by the success of agricultural production, which must be developed in a more sustainable way. The improvement of health and recreation possibilities of coastal inhabitants has been identified as a high priority. This also includes a more effective use of mud for health care purposes. To satisfy all these needs the local population should be involved in managing natural resources (e.g. mud, fish stocks, etc.) and increasing their responsibility regarding the development of the area.

Indicators and actions necessary for resolving problems:

3.5.1 Actual participation of the local population in the selection of alternatives for the planning process and implementation of projects in the coastal zone

Public participation in the planning process is often insufficient. According to the 2003 Planning Act all interested parties and people must be involved in the planning process.

3.5.2 Those groups in society whose livelihoods are under pressure; improved living conditions adjacent to the bay; migration to or from the project area

Different pressures that affect living conditions in the area were identified, including pressures to the bay. One particular pressure that should be taken into account already in the planning process is a possible flooding in the area that occurs once every 30-40 years. The potential impacts of increased tourism on the living conditions of local inhabitants were evaluated especially regarding the need for dredging a channel for yachts and its impact on the bay's ecosystem.

Photo 5: Flooding in Haapsalu in January 2005

Photo by Tõnis Padu

3.5.3 *Intensity of water use*

The project evaluated the consumption of potable water by the population and possible trends in the future. The pricing of water must take into account the ability of water consumers to pay for the water, because all people have a right to consume high quality drinking water.

3.5.4 *Quality of therapeutic mud*

The maintenance of the quality of mud and its availability can be in conflict with other targets in the area, e.g. with maintaining sustainability of agriculture and industry in the area and enhancing tourism. The annual excavation of mud (400 tons) is not high compared to the estimated amount of active mineral deposit (162.300 tons). The areas of active mud deposits are surrounded by a protection belt of 1,000 m where economic activities including construction works as well as transport by motor boats, setting up of ice routes and ice collection are prohibited by the order of the Minister of Environment. The total amount of mud in the bay is nearly 4 million tons. These deposits are not fully protected, which may cause deterioration of mud quality in the bay.

3.6 Developers

Regardless of the high land price in coastal areas there is a growing interest in building up these areas. Therefore, growing pressurez on the bay are evident. There are certain restrictions for construction of new buildings in coastal areas, e.g. 50 m from the water line in towns and 20 m in some areas in Haapsalu town. The development and construction works cause possible conflicts between local municipalities, population and developers. The importance of an improved planning process is evident, especially due to possible flooding in the coastal zone.

The lack of detailed plans causes uncontrolled changes in land-use in the project area, especially in coastal areas suitable for summer resorts. According to the Estonian Environmental Strategy 2010 one of the goals is to achieve and maintain good quality of coastal waters for the biota as well as for human needs.

4. Possible solutions and lessons learned

The main aim has been to demonstrate the conditions under which spatial conflict resolution in the case area can be handled using a bottom-up approach by covering all the problem dimensions. To meet this general objective comprehensive data about existing conflict sources, e.g. environmental quality of Haapsalu Bay and natural resources, was collected and evaluated. Many ongoing and planned projects in the region are aimed at handling environmental problems and economic development separately (usually in different sectors, e.g. tourism). The problem with this approach is that conflicting interests and goals cannot be handled, with a risk that results are jeopardised. In specific cases it is often obvious that there is a mix of basic facts and values, resulting in a struggle for a practical solution.

The information collected within the project will help to develop proposals for a more sustainable use of resources and possible conflict resolution. The case of the Haapsalu Bay area is quite typical and can be applied to other parts of Estonia and the other Baltic countries. Therefore, the methods for conflict resolution will have a much wider impact. In other words, the existing methods of system analysis and conflict resolution in coastal zones were used to handle the multi-dimensional aspects of sustainable development. To reach the goal the collected information was generalised and integrated to make it understandable for a wider audience in the process of developing solutions.

The second goal was to facilitate spatial panning through the development of practical, integrated methods, which are simple and built on the existing methods, such as Environmental Impact Assessment, Strategic Environmental Assessment, Risk Based Decision, etc. There are several serious threats that may be a hindrance to the improvement of the quality of life in the Baltic Sea Area. Two of the main threats are conflicting interests and goals, and environmental degradation. The environmental threats are also very complex and versatile. Several of the existing problems have historical roots, but the rapid growth of the economy will create new problems. There often is no systemic view in the handling of this problem. This is the case for example regarding investments in environmental technology. In decision-making there frequently is a lack of concrete approaches in specific cases and of scientifically validated methods. The results are therefore too general to be of great practical value. Therefore, proposals for a development plan were developed during the project period that could help Haapsalu municipality to achieve the status of an international resort, an environmentally friendly region and a suitable investment place.

Finally, an attempt to initiate a forum for scenario development and harmonisation of various interests in order to resolve potential conflicts in the future was made. The idea was to gather stakeholders representing different sectors and interests. This was implemented in the form of partner meetings and stakeholders' seminar.

5. Conclusions

Coastal planning and control measures for development are essential to protect the water quality of the bay. Recommendations for conflict handling were developed using the case study of the Haapsalu Bay area as an example and introduced to the stakeholders at seminars and roundtables. In the case of Haapsalu the goals of coastal development and economic growth should be balanced with environmental protection. It was demonstrated that organisational aspects of implementing EU water policy (political, research, administration, etc.) and problems of communication and information exchange between various levels of governance present major difficulties in the implementation of this policy. Involving multiple stakeholder groups in the development and implementation of EU and national water policies is of critical importance; however, this is not always feasible for various reasons. Traditionally, a major bottleneck in the implementation of environmental policies is created when experts produce highly technical information that becomes incomprehensible to non-experts. Therefore, proposals that are understandable and possible to implement were produced. When dealing with the coastal areas goals for coastal resource use and protection have to be established. This involves an inventory of coastal resources as the first step, as well as an analysis of existing problems and future needs.

The project enabled an increase in the level of experience and knowledge of the municipalities regarding European cooperation and EU environmental protection principles and practices. This experience will hopefully increase the level of sustainability of these municipalities when implementing future environmental programmes and in communicating on EU level including within the Pan Baltic cooperation.

Results from the study can help environmental authorities and decision and policy makers in assessing the effectiveness of measures taken to reduce pollution load on the Baltic Sea, in determining the relative significance of various pollution sources for the Sea, in evaluating the environmental impact of investments on the waste water treatment plant and consequently in establishing more realistic goals for the management of coastal areas.

6. References

Brandt, N., Gröndahl, F. & Loigu, E. (2003). Nutrient mass balance of the Haapsalu Bay, Baltic Sea, Estonia. 30th Pacem in Maribus. A year after Johannesburg. Ocean Governance and Sustainable Development: Ocean and Coasts – a Glimpse into the Future. Kiev, Ukraine, October 27-30, 224-234.

Haapsalu linna arengukava 2004-2009. Kinnitatud Haapsalu Linnavolikogu 29.10.2004 määrusega nr 42. 21 p.

Hood, M. & Iital, A. (2006). Sademevee reostuskoormuse uuringud USA-s ja Eestis. Keskkonnatehnika 2/06, 10-13.

Iital A., Gröndahl, F., Brandt, N., Loigu, E. & Reihan, A. Impact of changed nutrient fluxes to the water quality of Haapsalu Bay, the Baltic Sea (not published).

Statistikaamet (1997). Agriculture 1996. Yearbook, Statistical Office of Estonia, 130 p.

– (2005). Agriculture 2004. Yearbook, Statistical Office of Estonia, 80 p.

Klaipėda Deep-Sea Seaport Development

Olga Belous[1], Saulius Gulbinskas[1]

1. Introduction

Presently, the Lithuanian Coastal Zone (LCZ) is the most economically, socially and environmentally developing region in the country. Klaipėda is the only port in Lithuania. It is the Lithuanian transit centre connecting the main transportation corridors between the East and West via sea routes. A lot of new marine, industrial, mining as well as service organisations are concentrated in this unique natural landscape of coastal zone, including shipbuilding, ship repairing, stevedoring and shipping companies; chemical substance storages and oil mining. All of these are interconnected in the regional socio-economic system. In order to the improve system's efficiency and to expand East-West sea trade Lithuania has to develop the port of Klaipėda, maintaining its competitive ability

[1] Klaipėda University – Klaipėda – Lithuania.

with other Baltic Sea ports (The Law of Klaipėda State Sea Port 1996). The water area of the port is a natural channel connecting the Curonian Lagoon and Baltic Sea. The west bank of the channel is the coast of the Curonian Spit and is designated as a national park.

Negative environmental impacts of above mentioned objects could destroy both shore and open water natural ecosystems and lead to a reduction of the local flora and fauna biodiversity. This part of Lithuania, which its high impact of human activity, needs a sustainable development programme in order to save itself for future generations, and in particularly, such natural values as the Curonian Lagoon and Curonian Spit.

The aim of this study is to distinguish possible natural, social and economic conflicts of the establishment of the Klaipėda deep-sea port development and to propose recommendations and methods for preventing different coastal zone stakeholder conflicts, or find conflict resolutions using cross-border experience exchange.

In this case, a fulfilment of Strategic Impact Assessment (SIA) and Environmental Impact Assessment (EIA) is described. Due to the fact that the public is one main stakeholder, the importance of public participation in the procedure of real democratic evaluation is shown. The method of Integrated Coastal Zone Management using advanced conflict prevention and resolution means is proposed as a solution. The wide range of stakeholders as target groups of the study is presented. The different tools for stakeholder communication and education, such as training courses for various levels of decision making and involving independent advisers (the general public, NGO and scientist representative group) for advanced study of coastal zone conflicts are introduced.

2. Problem background

2.1 Importance of Klaipėda and its port for the economy of Lithuania

Klaipėda city was founded on the western coast of the Baltic Sea in 1252. It stands out among other cities of Lithuania due to its old and rich historical heritage and specific architecture that are common to the cities of Western Europe. Klaipėda is the third largest city according to the number of residents in Lithuania, with a population of approx. 200.000 people. The area of Klaipėda town is 84 km^2. It stretches 15 km along the Curonian Lagoon with a width of only 2-3 km.

Klaipėda distinguishes itself through fast economic growth. It is the second city in the country according to the economic potential. More than 15% of total foreign capital is invested in Klaipėda (Klaipėda county strategic development plan 2004). Klaipėda is the most rapidly growing industrial and transportation center not only in Lithuania but also in the Baltic countries. Port related companies flourish in Klaipėda. The geographical position of Klaipėda indicates tourism as the most prospective business in the city and region. A number of tourist companies offer their services in Klaipėda. Klaipėda is an ice-free port, a juncture where sea, land and railroads meet. A good land transportation infrastructure guarantees successful connection between the East and West.

Photo 1: Klaipėda city – the only port in Lithuania (Klaipėda city municipality 2006)

Klaipėda is a multipurpose, universal, deep-sea port, providing high quality services complying with the requirements of the European Union. 19 big stevedoring companies, ship repair and ship building yards operate within the port, and all marine business and cargo handling services are being offered. The annual port cargo handling capacity is up to 40 million tons (Klaipėda State Seaport 2006).

The average cargo handling capacity during year 1991-2004 ranged between 15-20 million tons of different cargo (Klaipėda State Seaport, 2006). In year 2005 there were 21.8 million tons of cargo handled in Klaipėda port (Klaipėda State Seaport 2006).

The port operates 24 hours per day, 7 days a week all year round.

Klaipėda region has been developing fast in recent years with the Klaipėda port development greatly contributing. The construction of the ferry port, the construction of the oil terminal at Butinge meant that Klaipėda region has an economic importance for the development of the whole country, but with this also came the matter of coastal zone management issues. One more interest in the Baltic coastal zone is oil extraction on shore as well as off shore. These issues will be described in the next chapters.

Fig. 1: The capability of Klaipėda port 1991-2004 m., mio tons

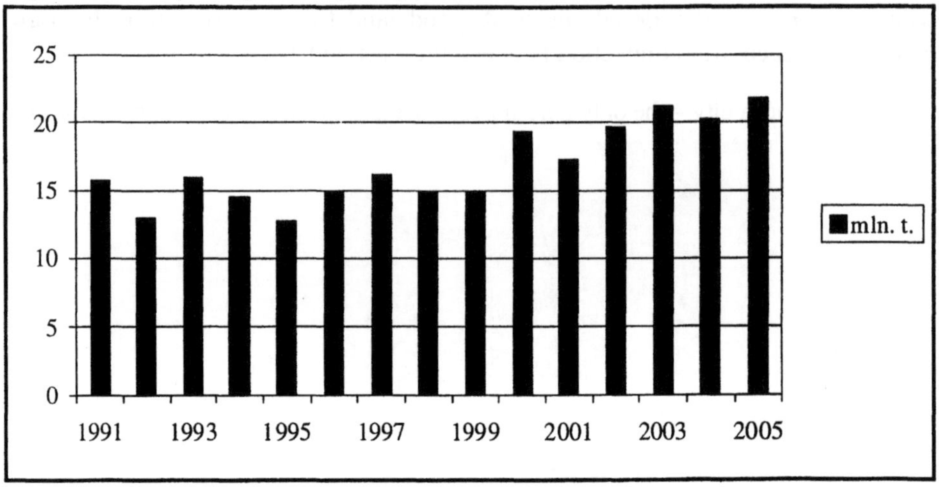

Klaipėda State Seaport, 2006

2.2 Chronological overview of coastal zone issues in Lithuania

2.2.1 Construction of the international ferry port 1981-1986

During the time of construction Lithuania was part of the Soviet Union. In 1980 the Solidarity movement was active in Poland, the situation became uncontrolled, and a state of emergency was declared. Threat to railway transit from the Soviet Union rose, especially for military going through Poland to the German Democratic Republic. Moscow decided to construct a sea ferry connecting Klaipėda and Mukran (the ferry port at Sassnitz, Germany).

The southern part of the port had to be deepened and new land mass had to be formed. Celebrities and members of a growing green movement proposed that it would negatively affect the unique nature of the Curonian Spit and endangered plants growing on the coast of the Lagoon. Lithuanian and Russian scientists were concerned about potential changes to the Curonian Lagoon ecosystem and protection of fish resources.

A mitigation mechanism was used when making the decision to build the ferry, committing money for construction of new fish farms. To evaluate the impact of ferry construction and exploitation on the Curonian Lagoon ecosystem and determine the damage to the fishery, an extensive scientific research programme was created. Every Lithuanian Science Institute involved with nature conservation was engaged in the implementation of this programme. To protect endangered plants in the Smeltes peninsula close to the newly build international ferry port a botanical reserve was build. Damage to the fishery then was estimated as

8 million roubles. Due to the changed political situation, compensation was never paid (Goss 1990; Safronovas 2002).

2.2.2 Construction of new oil terminal in 1988-1999

In the late 1980's Klaipėda oil terminal needed to be reconstructed or a completely new terminal needed to be build due to unsustainable technical and environmental features. Soil and groundwater in the terminal territory was strongly polluted by oil products. During a storm in autumn of 1981 the tanker „Globe Assimi" crashed while setting sail from the port of Klaipėda. 19000 tons of fuel oil were spilled. The port entrance zone was not safe for oil loading. Various new locations for terminal construction were discussed. The main alternatives were: Klaipėda or Melnrage, Karkle, Butinge – small seaside towns north of Klaipėda. The necessity to build a new oil terminal arose in 1991, when Lithuania regained independence and sought for oil supply from alternative oil resources avoiding potebntial blockades from Russia.

The conflict between different coastal zone exploitation and business interests starts to show. It was tried at the time to form an opinion about the necessity of oil transportation in Lithuania and if the construction of a new terminal was needed. It was thought that priority should be given to the development of recreation and tourism, because these would provide larger economic benefits and preserve unpolluted nature, and that Lithuania's oil supply and oil transit should be developed through foreign, especially Latvian terminals. This idea was especially supported by the green movement, the Latvian government and social organisations.

To find an optimal location for building the new terminal, the Lithuanian Government set up three groups of scientists, which analysed local alternatives from ecological, economical and social aspects and gave their conclusions. With reference to these, the Lithuanian Government decided to construct a new oil terminal in Butinge. Independent foreign experts (Det Norske Veritas), as well as Lithuanian and Latvian scientists carried out the environmental impact assessment. Lithuania and Latvia signed a damage compensation agreement, in case the environment would be polluted. Butinge terminal implemented an extensive monitoring programme. Every year results from this programme are published for interested persons and the general public.

Results of the environmental monitoring of Butinge oil terminal show that terminal operations generally do not influence the conditions of the sea environment. But there were a number of incidents when oil was spilled at sea. However, it has not been established how to compensate the damage experienced by other marine resource users in those cases. Lithuanian fishermen lost valuable fishing territories when the new oil terminal was build, and they have not been compensated (Lazauskienė et al. 1995; Gulbinskas et al. 2006).

2.2.3 Oil extraction in the Baltic sea 1984-2004

A few decades ago the company „Petrobalt", a joint venture between the Soviet Union, Poland and the German Democratic Republic began looking for oil in the southern part of the Baltic Sea. Potential structures were found, and. oil extraction was started in the German and Polish economic zones.

Two decades ago oil was discovered and its resources were evaluated in the nearshore structure D-6, opposite the Curonian Spit. An incident occurred during extraction leading to oil pollution of the sea and endangering the Curonian Spit coasts. But despite this accident it was decided to exploit the bore. Preliminary work on oil platform construction was started, supervised by the Soviet Union Oil Ministry.

An accident in the D-6 structure raised specific questions: is oil extraction in the sea sufficiently safe, will modern environmental protection technologies be applied, and will this guarantee prevention of accidents, what will the outcomes be of possible accidents to the marine ecosystem, as well as the public opposition, the Lithuanian Government and responsible environmental protection organisations asked Lithuanian scientists under supervision of the Science Academy to perform an expert evaluation of this project. Other Soviet Union Science Institutes also took part in this evaluation. Conclusions of the experts were not favorable for oil mining. Moscow decided to stop preliminary work (Exploration & Production: Offshore Technology Briefing 2006).

2.2.4 New situation: from 2002 until now

After confirmation of the Russian sea border, the oil-well D-6 was situated in the Russian economic zone and the company „Lukoil" decided to start oil extraction.

The D-6 structure or Kravtsovskoje oil-well is very close to the Lithuanian sea border, opposite of the UNESCO protected Curonian Spit. Official Lithuanian institutions and nongovernmental environmental protection organisations object to oil extraction in this place. Despite all objections from the Lithuanian side, „Lukoil" started oil extraction in 2004, because its activity is legal in Russia. But Lithuanian and Russian legal systems and international environmental impact evaluation liability differ. Lithuania is looking for international support, demanding independent environmental impact assessment, risk analysis and the signing of an interstate agreement regarding accident outcomes liquidation, damage compensation and environmental monitoring. Russian official institutions have offered to solve disputes regarding damage compensation in case of accidents according to international law; however they will not be signing any preliminary agreement or take responsibility for any preliminary liabilities.

Other oil resources also being explored in the Lithuanian economic zone, but Lithuania is the only Baltic Sea state, which did not plan to and has not exploited its oil resources. Furthermore, official Lithuanian institutions and international organisations offered to declare a moratorium for oil extraction in the entire Baltic Sea, as it is deemed environmentally sensitive.

In 2005 Lithuania and the Russian Federation decided to form expert groups and organised a sub-study on environmental impact assessment. As a result conclusions regarding oil extraction prevention measures, which could reduce extraction risk, as well as cooperation between the two countries, were presented. The necessity to sign an agreement between Lithuania and Russia was emphasised (After design evaluation ..., 2006).

2.2.5 Coastal erosion

Between all Baltic Sea states Lithuania has the shortest coast line, thus wise use and protection of the coastal zone are essential. Due to global warming (rise in sea level, increase in storm activity, decline of sand resources vital for coast formation processes) and impact of human activity (hydro technical constructions, stabilisation of sand dunes, deepening of Klaipėda port and recreational activity etc.) coastal erosion intensified. This problem became especially acute for the Lithuania's largest resort, Palanga, where over the last 5 years the beach was almost totally washed off creating a critical situation during the summer season, when the highest numbers of holidaymakers appear.

Although the scientific community already gave such a prognosis 40 years ago and all that time offered to apply coastal protection prevention measures, government institutions did not pay attention and left coastal protection to enthusiasts and local authorities. No material resources were given. Coastal resource users' activity was not coordinated, and the building of hydro technical constructions was performed without scientific recommendations.

The Lithuanian Environment Ministry published a coastal management strategy in 2001, and a programme for coastal protection in 2003. It is trying to allocate finances in the state budget and access EU funds for the implementation of various projects. The Klaipėda County Head Administration is responsible for implementing coastal protection programmes. But the law and relevant institutions are still insufficiently prepared for any implementation of integrated coastal zone management. There are no qualified coastal zone managers, or even any such positions in the relevant institutions (Žaromskis 2006).

2.3 Main measures of the port development

Being the only sea port in Lithuania, Klaipėda port is of strategic value for the state. In 1994 port reconstruction and development started to increase the port's competitiveness with other states of the Baltic Sea. The main measures for port development include:

- deepening of water area

- reconstruction of the port entrance
- reconstruction and building of embankment
- construction of new terminals and development of port infrastructure.

All these port activities cause a number of specific problems described later in this study (Jokšas et al. 2003; Lazauskienė et al. 2000).

2.3.1 Dredging

Dredging and increase of navigation channel depth leads to environmental changes, especially on the water and sediment exchange between the Curonian Lagoon and the Baltic Sea, the ecosystem overall, as well as fish migration and fishing conditions. These changes will be monitored on a permanent basis through implementation of the environmental monitoring programme of Klaipėda port.

Photo 2: Klaipėda port waterscape

Klaipėda State Seaport, 2006

2.3.2 Dredged material disposal and dumping

The amount of dredged material reaches up to 1.5 million m^3/a (total initial and maintenance dredging) at Klaipėda port. The problem of reusing dredged material has not been solved. Sections of the material are also polluted. Soil dredged in the port is mostly disposed of through dumping at sea except for highly polluted material. Over a long period of time material dredged from the port entrance channel have been dumped in this specific area and have subsequently been removed with the long shore drift. This in turn has stimulated coastal erosion.

The deepening activity is interrupting fish migrations and the dumping is interrupting fishing activity and leading to the loss of fishing grounds at sea.

The Environment Ministry has approved a regulation for soil dredging and disposal at sea, which provides for an evaluation of soil pollution and acceptable pollution limits. It also includes a ban on dumping of polluted soils at sea. Construction of a special storage yard for the polluted material is planned for Klaipėda port. The design of this yard should be started in 2007.

The Environment Ministry is also currently preparing a compensation order for the damage to fish stocks.

2.3.3 The reconstruction of the port entrance

Piers were lengthened, their configuration was changed and the mouth channel was narrowed and deepened to 14.5 m during the port entrance reconstruction. This changed the distribution of out flowing water from the Curonian Lagoon to the Baltic Sea and hydrological, sedimentation and ecological processes associated with it. The port entrance channel interrupts the long shore sediments drift in S-N direction and creates a sand deficit along the northern coastal slope. In the coastal zone sectors next to the port lithodynamical processes have intensified, and bottom relief and coastal line transformation progressed.

After conducting the environmental impact assessment it was decided to use sand dredged in the port entrance channel for near shore nourishment. During 2003-2005 approximately 5000 $m^3$of sand was dumped near the shore. This experiment showed very good results in reducing coastal erosion.

2.3.4 Port territory and infrastructure development

For further development of the port its geographical location is very important. The area of the port of Klaipėda stretches like a narrow belt along the strait of Klaipėda. It is separated from the remaining continental part by the city of Klaipėda. The water area of the port borders on the Curonian Spit, which is included in the UNESCO World Heritage List as valuable cultural landscape. Additional areas for construction of the new port terminals and loading storage are needed. Opportunities of increasing navigation intensity are limited by the narrow, one-way mouth channel and the port's fairway. These conditions have led to an increased interest in analysing alternatives (Gripaios 1999).

For port service additional land areas outside the port management area are needed. This land is under the administration of Klaipėda city and Klaipėda country. Out of this conflicts arise regarding environmental protection in relation to opportunities for the use of sensitive areas, as well as regarding economical and social conflicts in relation to the purpose of use of the area.

Klaipėda city, port and county have arranged strategic environmental impact assessment of the suggested alternatives. Development of general plans of Klaipėda city and Klaipėda port have been started in 2003-2004 and should be finished in 2007.

3. Conflict description

3.1 Perspectives of Klaipėda port

Klaipėda port extends on a north-south axis along a narrow band over 10 km in length lying mostly adjacent to the urban area of Klaipėda city. The water area of the port is a natural channel connecting the Curonian Lagoon with the Baltic Sea. The west bank of the channel is the coast of the Curonian Spit and is designated as a national park. These site conditions place physical constrains on any large-scale expansion inside the port.

Klaipėda port capability is 40 mio tons per year putting it in fifth place after the biggest ports in the Baltic, such as Ventspils (80,2 mio tons), St.Petersburg (50 mio tons), Primorsk, and Tallinn (47 mio tons). The total cargo handling volume of Klaipėda port reached 20 mio tons after 2003.

To meet the cargo traffic demands up to the year 2025, three port development scenarios have been envisaged, including (JICA 2004):

- Scenario1 : existing port development only
- Scenario 2: inner port expansion
- Scenario 3: outer port expansion.

Even if the existing port is fully renovated, the traffic demand would exceed the existing port capacity by around 2015 to 2017. To cope with the anticipated capacity shortage and to meet the shipping needs based on Klaipėda port being able to receive Baltmax-type vessels, Klaipėda port should be expanded beyond the existing port territory (JICA 2004).

3.2 Key environmental features in and around the port development areas

Klaipėda port is very important for the economy of the country and a major influence on the life of the city. It is also located in an area of considerable environmental importance and sensitivity:

- The Curonian Spit in the west, which has internationally important landscapes, culture and ecology, and is designated as a National Park and World Heritage Site.
- The channel between the Spit and the Port, through which commercially exploited and rare fish, and internationally important birds migrate each year.

- The village of Melnrage and the Baltic coast in the north, which are used for recreation and tourism by local people and visitors.
- Klaipėda city in the east, where commercial and residential areas are located close to the port with little or no buffer in places;
- Land immediately to the south of the port is a groundwater protection zone as it contains boreholes from which water is extracted to supply the city.
- The Curonian Lagoon in the south is a fish spawning ground, a resting and overwintering site for birds, a tourist attraction and supports commercial fishing.

The port is therefore surrounded on all sides by areas and features that are of local, national and international importance, which are sensitive in different ways and to varying degrees to damage and disturbance. Proposals to expand the port therefore need to be developed and implemented with a great deal of sensitivity towards environmental considerations to prevent damage and disturbance to important assets.

3.3 Port development direction

One of the alternatives is to expand the port territory in southern direction on the East coast of the Curonian Lagoon. Port development in this part is limited by the III Klaipėda city water supply site, which provides drinkable water for 60% of Klaipėda's residents. An Alternative water supply source is only 50 km in the Silute direction. Port development to the South would lead to irreversible ecological changes. Port development would affect the coast of the Curonian Lagoon. This port development alternative would not solve the problem of maximizing navigation intensity.

In the south of the Port, many physical constraints exist to port expansion, including the existence of conservation areas for fresh water intakes and shortages. The study on the port development (JICA 2004) stated that port development to the South is not a viable solution.

Another port development alternative is port expansion to the North. This alternative suggests to develop an island in the sea opposite to Melnrage, and to create a deep-sea port. A study of this type of possible port construction has been arranged (JICA 2004).

The sea basin near the existing entrance channel and Melnrage beach can provide a large open space for outer port expansion. The required marine facilities for the outer port include (JICA 2004):

- Petroleum jetty (Baltmax-type)
- Grain terminal (Baltmax-type)
- Two fertiliser terminals (Panamax-type)
- General cargo terminal (Panamax-type)
- Container terminal (Panamax-type)
- Breakwaters.

Fig. 2: Outer deep-sea port location near Melnrage village

JICA 2004

To accommodate the above terminals an artificial island (1500m long and 700m wide) will be developed 350m offshore from the coast line. The reclaimed outer port area will have road/railway access linked to the existing railway and road networks inshore.

Construction of this type of port is not considered within the Klaipėda city strategic development. This would cause conflict with local people, land and realty owners. Construction of the deep-sea port would totally change coastal formation processes and would have a significant impact on other coastal zone resource users.

The port development committee has agreed with port development to the North. To analyse the present situation, a specification of the conflicts, the creation of an additional activity programme, and the implementation of integrated coastal zone management principles is needed. The Committee of Strategic Planning of the Lithuanian Government has approved this initiative.

3.4 Possible environmental impact during construction and operational phases

The location of the outer port is less than 1 km from the Curonian Spit, one of Lithuania's most important landscape, tourism and nature sites, and 600 m from the town of Melnrage, which has 1500 residents. Impacts of the construction phase can be described by the following (JICA 2004):

- Plumes of turbid water overflowing from the reclamation site could discourage people from using nearby beaches if these areas were affected in the summer.
- Oil, fuel and other chemicals handled on site could pollute adjacent beaches and sea areas if they were spilled.
- Fish stocks and biodiversity could decrease if the construction site prevents fish migration along the coast into the lagoon to bread.
- Any birds overwintering at the north of the Spit or nesting there in the spring could be disturbed by the noise or visibility of the construction activities.
- Melnrage could be considered a less desirable living area because of the presence of a large construction site, so the value of land and property may decrease.
- The site will be visible over a long length of coastline, which is heavily used for recreation by local people and visitors.

Impacts of the operational phase include:

- Chemical and physical pollution, cargoes to be handled in the new port would pollute the water if spilled.
- Limitation of sediment supply to the coastline to the north, so beaches could erode.
- Water quality will need to be high in the inshore bay to be suitable for recreation, and this could be difficult to maintain because water will only be exchanged slowly with seawater from outside due to very limited tidal action.
- Sediment accumulation at the mouth of the bay could make the recreation area less attractive and could reduce water quality by further limiting flushing.
- Residents in the south of Melnrage will be disturbed by noise from port related road and traffic on the new routes around the south of the village.

Negative impact on the landscape will come from converting an area of natural coast used for recreation by local people into a large industrial port. The outer port will be visible from Melnrage, from beaches to the north, and from the Curonian Spit, an area which has been designated as a National Park and UNESCO World Heritage Site because of its landscape beauty.

From the EIA point of view the development of the outer port in the village of Melnrage would be less desirable due to:

- Adverse impacts on surrounding natural landscape;
- Possible decrease of nearby property values;
- Likely beach erosion to the north of the outer port.

Visitors to the Curonian Spit make a significant contribution to the local economy, so there could be further negative impacts if numbers were to decline. Otherwise, by increasing trade and government revenue, new port could improve social and economic conditions throughout the country. A new port would also generate local employment, which could improve social conditions and stimulate the economy of Klaipėda.

4. Relevant stakeholders and conflict resolution methods

4.1 Categories of stakeholders

To ensure successful development of the project it is very important to determine stakeholders and define collaboration between these. In Figure 3 the main categories of stakeholders are described.

Fig. 3: Coastal zone stakeholders

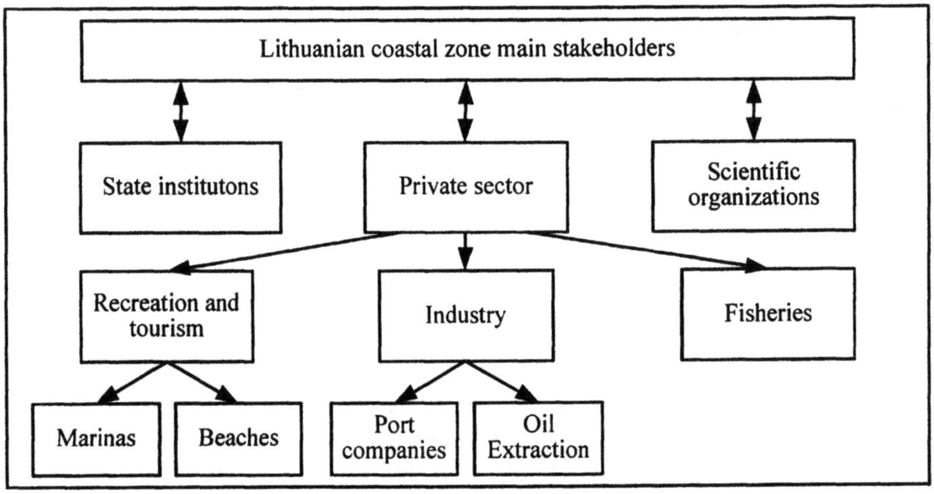

Belous and Gulbinskas, 2005

The development of the coastal zone (National Sustainable Development Strategy, 2003; The Law of Klaipėda State Sea Port, 1996) is of great concern to state institutions such as:

- The Government of the Republic of Lithuania;
- Ministry of Transport and Communications of the Republic of Lithuania;
- Klaipėda state Sea Port Authority.

Institutions such as the Ministry of Environment, Klaipėda County Head Administration, Klaipėda Municipality, Klaipėda Public Health Center, Lithuanian Maritime Safety Administration and Culture Heritage Protection Department are responsible for the District Development General Plan and for coordinating the territorial planning and fulfillment of Strategic Impact Assessment and Environmental Impact Assessment (Coastal Zone Protection Programme 2003; Regulation 2000; The Law on Environmental Impact Assessment 2000).

Other interested parties include:
- Curonian Spit National Park Directorate;
- Seaside Regional Park Directorate;
- Coastal and Curonian Lagoon fishermen;
- The general public (for example citizens of Melnrage village, NGO).

4.2 Decision making process and potential conflict prevention

By the Decree of the Government of Lithuania (2003, December) a plan of measures for the implementation of a master plan for the territory of Lithuania was approved including the provisions for investigating the possibilities of the construction of Klaipėda deep-sea harbour. The study of Japan International Cooperation Agency (JICA) determined that cargo turnover by 2015-2017 would exceed the present capacities of the port. In order to remain a competitive port, Klaipėda port should be expanded beyond present limits. The JICA study investigated 4 port development alternatives. When the analysis of the alternatives was made it was determined that it would be expedient to develop the port in the Baltic Sea near Melnrage because with the remaining investigated alternatives it would be impossible to reach the necessary depth of 17-18 meters for such a port (JICA 2004).

It is a well-known fact, that with proper and early planning, the support of policy makers, and a relatively small investment, a rigorous evaluation can be very powerful in assessing the appropriateness and effectiveness of programmes (Baker 2000).

In accordance with the „Inventory of order of assessment of strategic consequences on environment of programs and plans" Klaipėda State Sea Authority organised the assessment procedures.

Correct and comprehensive information relating to potential problems regarding port development issues gives decision makers opportunities to solve these in time and prevent future conflicts (Mayoux 2002).

The forecast of possible issues and its consequences for harbour development and methods of assessment were determined. Essential environmental and social sensitive areas were identified, which and are represented in Table 1.

The main impact for which no compensation is possible by any measures is the impact on landscape. The construction of an industrial object is planned close to UNESCO cultural heritage object and popular recreation areas. However, port development necessity is determined by positive economical growth and importance for the whole country.

Tab. 1: Scope of Strategic Impact Assessment

Environment and social components	Possible impacts	Content of data for the assessment of the state of environment and prognosis
Water	Changes of hydrochemical and hydrological regime due to the construction of the port and economic activities	Statistical summary of long-lasting years observations of sea water quality, hydrodynamic coastal regime and changes after port construction
	Additional pollution and increased risk to pollute water territory during accidents	Hydrodynamic regime of coastal zone and prognosis of water quality
Bottom sediments	Changes of sedimentation processes	Composition of bottom sediments and their contamination
Landscape	Coast erosion	Investigations of sediment transport and coastal dynamics Prognosis of changes of coastline, beaches and dunes
	Degradation of natural landscape due to the development of port	Recent landuse and its changes
Geological conditions and seabed	Changes of geological processes	Engineering-geological investigations
Biodiversity	Biological resources, important and protected species and NATURA 2000 places	State of natural habitations and assessment of changes Bird migration and their wintering places Fish communities, spawning conditions and migration. Assessment of NATURA 2000 territories
Cultural heritage	Loss of objects of cultural heritage	Underwater archaeological objects
Public health	Changes in Public Health	Determination of significant direct and indirect consequences on public health
Social economic environment	Increase of level of noise due to additional transport	Assessment of the level of noise and possible protection measures. Analysis of traffic alternatives.
	Loss of attractiveness of recreation territory. Loss of the value of real property.	Assessment of the conditions of recreation and prognosis of possible changes.
Air quality	Air pollution	Impact of port, ships, land transport on the air quality

Source: *Klaipėda Deep-sea Port Strategic Environmental Impact Assessment, 2006*

Information about the strategic impact on the environment will be given in the assessment report following the Decree of the Government of Lithuania (Valstybės žinios 2004), which approved the „Inventory of order of assessment of strategic consequences on environment of programmes and plans".

The benefits of the assessment of strategic consequences on the environment include the following: determination, description and assessment of possible significant consequences on the environment from the implementation of Klaipėda deep-sea harbour; confirmation that there will be consultation with relative state and municipal institutions, the general public, taking into consideration the transparency of such consultations that assure the results of procedures; affiliation that the organisers of Klaipėda master plan will acquire comprehensive and reliable information about possible significant environment impacts of the implementation of Klaipėda deep-sea harbour and will take these into consideration in their decision making process.

Klaipėda municipality organised a questionnaire related to the social aspects of the necessity of a deep-sea port. Public opinion on this question is very important for the politicians of Klaipėda municipality who will be making the decision on the deep-sea port construction.

5. Possible solutions and lessons learned

5.1 Application of strategic impact assessment's method

Strategic Impact Assessment (SIA) may be used to strengthen the appraisal and evaluation of the Klaipėda master plan regarding thedeep-sea port development.

The effectiveness of SIA, when applied to policies, plans or programmes for enterprise development, depends as much on the effectiveness of its assessment process as it does on the quality of data, assessment methods and analysis, which it uses (Lee 2002).

In this case study, besides the information that is usually used to describe economic activities and construction of port developments and its consequences, the following data will improve the assessment of the state of the natural and social environment of the Klaipėda deep-sea harbour:

1. Bathymetric plan M 1: 25000 (2004) of the middle part of the Baltic Sea with the approaches of Klaipėda port;
2. Monitoring data of Klaipėda state seaport from 1995 to 2004 according to the programme approved by the Environment Protection Department of Klaipėda region.
3. Information accumulated during preparation of the report by Japan International Cooperation Agency (JICA 2004).

Consequences of constructing the deep-sea harbour will be determined by using hydraulic coastal digital modelling. The changes of the structure of flow and sediment transport will be obtained by comparing the parameters of sea flow regime with a „0" alternative. When choosing the planned layout of breakwaters in the model it is possible to minimise the consequences of the port for the coastal zone.

Zoning of water territory according to the contamination of bottom soils is carried out using geographical information systems. Bottom sediments and soils will be investigated in the marine geo-chemical laboratory certified by the Institute of Geology and Geography. The valuables of cultural heritage will be identified during underwater archaeological investigations: sunken ships, sheet piles, ship equipment etc.

In the NATURA-2000 area adjacent to the deep-sea harbour an investigation into biocenosis, the places of fish spawning and migration, bird accumulation places will take place. In addition, a summary of former observations and investigations from expeditions in 2006 will be made, as well as suggestions for the proposed Klaipėda state seaport monitoring programme (Paulauskas 2004).

The general public will be informed with the solutions of Klaipėda master plan, its alternatives and possible consequences.

5.2 Management of environmental aspects of deep-sea port development

Predictions of the impacts of the deep-sea port development will allow the organisers of economic activities and decision makers to prepare solutions and tools for the management of the environmental aspects of the port development (Tab. 2).

Information about strategic impacts on the environment will be detailed in the assessment report following the Decree of the Government of Lithuania (Valstybės žinios, 2004), which includes the approval of an „Inventory of order of assessment of strategic consequences on environment of programmes and plans".

The benefits of the assessment of strategic impacts on the environment include the following: determination, description and assessment of possible significant impacts on the environment from the implementation of the Klaipėda deep-sea harbour; confirmation that there will be consultation with relative state and municipal institutions and the general public, taking into consideration the transparency of such consultations that assure the results of procedures; assurance that the organisers of Klaipėda master plan will acquire comprehensive and reliable information regarding possible significant implementation consequences of the construction of Klaipėda deep-sea harbour on the environment and will take these into consideration during their decision making

Tab. 2: Consequences of deep-sea port development and its management

Environmental aspects	Evaluation	Equipment, installations
Hydrodynamic structure of the flow of the Baltic Sea	Statistic data processing of many year hydrometeorological observations Hydraulic digital modelling	Programme package „Statistics", two dimensional modelling system MIKE 21 of hydrodynamic phenomena.
Sedimentation and lithological processes in the coastal zone	Statistic data processing of environmental monitoring Laboratory tests of bottom sediments and sludge Bathymetry photos Statistical data processing of environmental monitoring Modelling of sediment transport	Acoustic depth finder, sediment catchers, laboratory equipment for soil investigations, modelling system MIKE 21
Variation of biological parameters	Investigation of biocenosis Investigation of fish spawning and migration Catching of hydrobiota Use of GIS for data processing	Peterseno and Van-Vino methods GIS, dredger, nets,
Coastal shores	Visual observations. Statistic data processing of monitoring. Underwater archaeological investigations.	Topographic and bathymetric photos
Socio-economic environment	Assessment of changing in the living environment Assessment of real property. Summary of enquiry of inhabitants.	

Source: Klaipėda deep-sea port strategic environmental impact assessment, 2006

6. Conclusions

- Strategic Environmental Impact Assesment and Risk Assessment using cross-border experience exchange are important instruments for harmonisation of planning in the costal zone in Lithuania.

- For a successful deep-sea port development efforts should be made in preparatory works regarding the port development project, considering the city's needs and consulting with environmental organisations as well as with local community.
- To ensure the development of the out port, Klaipėda State Seaport Authority should develop environmental procedures, as recommended in EIA studies. The level of environmental tools must be accompanied by a general city plan.
- The procedures prescribed by law are not always sufficient to solve all existing conflicts between different stakeholders and interest groups.
- The Lithuanian case study realised within the „Coastman" project shows that impacts on the natural environment can be minimised by various mitigation/compensation measures.
- To take into account different stakeholders' economic and social needs could help to debelop a compromise between these.
- The application of the mechanism of compensations could prevent conflicts between different stakeholders.
- In order to improve the quality of decision making educational courses on coastal zone conflict resolution should be offered for all stakeholders.
- Independent advisers including citizens, members of NGO and scientists should be involved in an advanced study on coastal zone conflicts.
- The explanation of the benefits of a deep-sea port development to the public should be carried out before any SIA or EIA.
- Contacts with all EIA stakeholders and the general public has to be made at the early stage so that all parties are informed and can participate in the process from the beginning.
- Project based methods should be used for the management of possible natural and social conflicts in the coastal zone. This could be a useful tool for better understanding and cooperation between interested partners.
- The successful further development of the deep-sea port design is possible.

7. References

After design evaluation of ecological impact of oil extraction in the D-6 structure on Curonian Spit environmental conditions. Report of Lithuanian and Russian experts (2006).

Baker, J.L. (2000). Evaluating the Impacts of Development Projects on Poverty: The World Bank, Washington DC, p. vi.

Belous, O. & Gulbinskas, S. (2005). Power Point Presentation for „Coastman" meeting "Legal, Organization and Economical Issues of Lithuanian Coastal Zone Management", Stockholm, April, 2005.

Coastal zone protection program (2003). Environmental Ministry of Republic of Lithuania; 2003-09-01 No. 44.

Exploration & Production: Offshore Technology Briefing (2006). www.touchbriefings.com/cdps/cditem.cfm?nid=1952&cid=10.

Goss, R.O. (1990). Economic policies and seaports. The economic functions of ports, Maritime Policy and Management, Vol. 17, 207-220.

Gripaios, P. (1999). Ports and their influence on local economies – a UK perspective: The Dock Harbour Authority, Vol. 79: 894, March/April 99.

Gulbinskas, S., Gasiūnaitė, Z.R. & Balsys, V. (eds.) (2006). The state of the Environment of the AB „Mažeikių nafta" Būtingė oil Terminal in 2000-2005, 45 p.

Japan International Cooperation Agency (JICA) (2004). The study on the port development project in the Republic of Lithuania, Final report, Vol. I.

– (2004). The study on the port development project in the Republic of Lithuania, Final report, Vol. II.

Jokšas, K., Galkus, A. & Stakėnienė, R. (2003). The only Lithuanian sea port and its environment Vilnius, 314 p.

Klaipėda city municipality, www.klaipeda.lt/klaipeda/selectPage (2006).

Klaipėda county strategic development plan (2004). www.stat.gov.lt/uploads/ docs/Darn_vyst_2006.doc?PHPSESSID=74813da4ed5d16f79a8bdd985c12bc 00.

Klaipėda State Seaport (2006). www.portofklaipeda.lt/ en.php/ general_information/ presentation_of_the_port/267.

Lazauskienė, L.L., Vaitonis, G. & Draugelis, A. (compiled by) (2000). Klaipėda port: economy and ecology Baltic Eco, Vilnius, 150 p.

Lazauskienė, L.L., Virbickas, J. & Gedvilas, V. (compiled by) (1995). The Butinge oil terminal (ecological state). Baltic Eco Vilnius, 166 p.

Lee, N. (2002). Strategic impact assessment and enterprise development. Institute for Development Policy and Management, University of Manchester.

Mayoux, L. (2002). From Impact Assessment to Sustainable Strategic Learning. Available at (last viewed on 21.12.2006): www.enterprise-impact.org. uk/ word-files/FromIAtoSustainableStrategicLearning.doc

National sustainable development strategy (2003). No. 1160: Available at: www.stat.gov.lt/uploads/metr_2005/print/doc_pratarme.doc (last viewed on 21.12.2006).

Paulauskas, V. (2004). Uosto terminalų palnavimas, Klaipėda: KU leidykla.

Regulation concerning sea coast protection and use (2000). Environmental Ministry of Republic of Lithuania, 2000-02-24, Nr. 73.

Safronovas, V. (2002). Klaipėdos miesto istorinės raidos bruožai, Klaipėda, 2002 (ISBN 9986310636).

The document of Klaipėda deep-sea port strategic environmental impact assessment's comprehension determination. LEI (2006).

The Law of Klaipėda State Sea Port (1996). Vilnius, No. I-1340. Available at: www.portofklaipeda.lt/en.php/port_regulations/port_act/555 (last viewed on 21.12.2006).

The Law on Environmental Impact Assessment of the Proposed Economic Activity of the Republic of Lithuania (2000). 18-04-2000, No. VIII-1636. Available at: ww3.ima.kth.se/Nordenvt20/Report2.doc (last viewed on 21.12.2006).

Valstybės žinios, Vilnius (2004). No 130-4650.

Žaromskis, R. (2006). Lithuanian coast and dunes. Klaipėda, 48 p.

Ecological and Sociological Studies as a Crucial Tool for Decision-Making in Primorsk Coastal Zone Management

Elena V. Victorova[1], M.B. Shilin[1], N.L. Plink[2], A.G. Kosheleva[3], G.G. Gogoberidze[2]

Photo: E.Viktorova

Abstract

The aim of this case study is to foster sustainable development in the Primorsk area by using environmental and sociological research for improvement of the decision-making process. The case is based on the Integrated Coastal Zone Management (ICZM) approach taking into account environmental safety and social stability.

1 Saint-Petersburg State Polytechnical University, Project Coordinator of the Russian Team, Saint Petersburg, Russia.

2 Russian State Hydrometeorological University.

3 Local Administration of Primorsk City Municipality.

1. Introduction

The case study in Primorsk was carried out within the framework of the Tacis CBC Project „Decision making tools for implementation of Integrated Coastal Zone Management (ICZM) in the Primorsk Area".

The overall case objective is to facilitate sustainable development of the coastal areas around existing and planned oil terminals by improving the quality of the decision-making process at local authority level. The realisation of this project will enable participants to develop and substantiate methods of support of decision-making processes in the field of coastal zone management on the basis of integrated monitoring of ecosystems and sociological surveys.

Based on the system approach ICZM defines the actions to be undertaken depending on the general purposes, priorities and problems of strategic development of the coastal area in the context of achievement of the best results based on the balance of economic, political and social interests.

Thus, the first aim of the case is to develop recommendations for the authorities to increase local public awareness and to improve the mechanism of public participation in the communication and decision-making process. The results obtained by processing the data collected during the conducted public opinion poll have been forwarded to Primorsk Municipality.

Ecological information has been collected and analysed. Ecological sensitivity maps developed as a result of data processing have been forwarded to the local authorities.

Oil contamination is the most important ecological problem of the Gulf of Finland and the Baltic Sea as a whole. Non-contact (remote monitoring) methods have been used to assess the degree of sea surface contamination by oil products, and reports of these studies have been published (3, 7, and 14).

2. Background information

2.1 Historical and geographical background

The town of Primorsk is located on the Karelian Isthmus at the distance of 148 km from St. Petersburg, and 43 km from Vyborg. Primorsk stretches along the eastern coast of the Strait of Bjorkesund over a length of more than 10 km. The Starit of Bjorkesund separates the Birch archipelago („Berezovye ostrova") consisting of several dozens big and small islands from the Karelian Isthmus. Primorsk is well protected against winds and waves and has a good harbour. Piers and quays for fishing boats, passenger ferries, yachts and other craft are located here. Primorsk is conveniently located for transport, and the city is crossed by a highway and railway.

Primorsk originated from a trading settlement of Slavic people from Novgorod, which was named Berezovy („Birch"). It was first mentioned in the annals in 1268. Later, the place was renamed Bjorke by the Swedish, and Koivisto by the Finns, which means „a place with birches". In 1940, according to the peace treaty between the USSR and Finland, the border line on the Karelian isthmus was moved 150 km north from Leningrad, and Koivisto became Russian. On 1.10.1948 the town was renamed Primorsk.

2.2 Natural and cultural assets

Today Primorsk is a small town with approximately 7000 inhabitants whereas in summer season the population can reach 10 000. Primorsk has 35 streets and passages, with the majority located along the gulf and the highway. The city is surrounded by pine forests, rich in mushrooms and berries. In addition, the city has many green plantations of linden, oak, alder, larch, ash, manchurian nut, lilac and fruit trees. The climate is a mixture of continental and marine. In winter the weather is changeable, and frequent cyclons bring strong winds and thawing weather, melting the snow even in December-January. The warmest summer month is July.

Primorsk's cultural and historical places are very popular with the local population and visitors. The most important one is Kirch – a Lutheran church designed by the architect J. Stenbeck was built in 1904 (Fig. 1). In 1992 the monument „Sail and Cross", designed by the Finnish sculptor A.Saalo, was erected. The cross is a symbol in memory of those who perished here in World War 2: military and civilian Russian and Finnish people and the sail is a symbol of friendship between Russian and Finnish people. There is also a Memorial – the common grave of the Soviet soldiers who took Primorsk. The names of the Finnish soldiers who defended Primorsk are written on a memorial plate placed on the old and half-ruined fort wall, situated on an island, the biggest among the „Berezovye Islands".

Currently, a large majority of Primorsk's inhabitants work at the joint stock company „Petrotral" and the Corporation „Primorsky Fisherman", as well as the Corporation „Primorsky Fish Factory", following the historical traditional specialization of the city in fishing and fish-processing. Primorsk forests are supervised by the employees of the Primorsk Forestry service. Primorsk is also home to the Research Center of the Space Missiles Corporation „Energy", well known both in Russia and abroad, and other enterprises.

The protection area „Berezovye Ostrova" („Birch Islands") is an especially valuable natural object in close proximity of Primorsk, established to protect its original landscape, a combination of pine and spruce forests, low and upper swamps, as well as a large number of migrating and nesting bird species. Since 1976 it is a part of the Vyborg protected area. The independent protected area „Birch Islands" was established by the order of the Leningrad Oblast Government dated 26 December 1996. The size of the protected area is 12000 hectares, including 7000 hectares of sea surface. The Birch Islands are covered with forests. They mainly consist of pine, spruce and birch. Also, alder, aspen and mountain ash can be found. In places of former settlements it is possible to find maples, oaks and other tree species. The herbs of more than 120 species are growing here. The Birch Islands are home to moose, foxes, white hares, minks, squirrels, and moles. As for amphibians and reptiles, the protected area harbours tritons, toad, frogs, adder and, rare for this region, the grass snake. In shallow-water bays a lot of fish, including commercial species, spawn, e.g. pike, pike-perch, bream, perch, white-fish, and in the brooks – trout.

Photo 1: Lutheran Church built in 1904 was designed by the architect J. Stenbeck

Photo by E. Victorova, July 2006

The most diverse in the protected area are the birds. In central parts of the islands there are chaffinches, bullfinches, orioles, some types of peckers, and others. In wild places it is still possible to observe a wood grouse and a blackcock. In the wetlands, on the overgrown sylvan lakes and on the littoral there is a real bird's kingdom. Grey seagulls and terns are nesting in colonies, broody hens and silver seagulls – in couples or small communities. Occasionally, on separate small stones jutting from the water the biggest seagull nests. Among the colonies of seagulls there are always nests of different diving and river ducks. About 110 bird species are nesting in the protected area and just as many arrive during seasonal migrations in spring and summer. Here is the breeding ground of some migratory birds, first of all swans, geese and ducks. This is why this protected area is also designated as a Ramsar site, especially protected under the International Ramsar Convention.

On the Birch Islands a number of bird species can be found, which are rare for the Leningrad Oblast region, e.g. sandpipers, polar and small terns, golden-eye duck, see eagle, fish hawk, osprey and others. In fact, the clean sea air, picturesque landscapes, heavy-bodied needle-leafed and leafed forests with meadows and swamps, and the bird's hubbub have made the Birch Islands one of the most remarkable places of Northwest Russia.

Besides the protection area „Birch Islands", there are other valuable natural objects in the vicinity of Primorsk that need protection, e.g. the littoral zone from Primorsk to the village „Manola", which is a resort area with light pine forests, the brooks Banny, Manolsky, and Karasevsky, which are spawning grounds for salmon, the artificial plantations of arbours in the city with centennial maples and linden, cedar, manchurian nut, as well as pine and spruce forests.

2.3 Primorsk Oil Port

The picturesque nature of the Primorsk area until recently attracted holiday-makers and tourists and, as it seemed, the future development of Primorsk was possible only in the form of a resort city. Nevertheless, in 2001 a large industrial object was built near Primorsk, which significantly tampered with nature. It was the special oil port „Primorsk“ located several kilometers southeast of the city, and the oil pipeline carrying the oil across the Karelian isthmus to this port (Fig. 1). Other types of anthropogenic impact in this region, such as the fish processing factory (dumping untreated sewage), sewage disposal installations, pollution of the water preservation zone of the village Banny with household sewage and sewage from the railway station, and also the felling of forests combined cannot cause the ecological damage comparable to the potential damage from this object, increasing with further development of the oil port complex, which may become the main anthropogenic pollutant in this region.

It is necessary to also note the political aspects of the ongoing construction. Before the start of operation of the terminal in Primorsk, Russia used to transport oil as transit good through ports in Lithuania, Latvia, Estonia and Finland. In total, up to 70 million tons (mt) of oil and processed oil were transported through ports in these countries. Actually, only ports in Kaliningrad and Saint Petersburg operated as genuine inner ports in the Baltic sea for Russian oil. As early as late 1990 the quantity of oil transported through these ports was about five million tons annually, from which slightly more than three million tons were transported through the Gulf of Finland, that is through Saint Petersburg.

Fig. 1: Eastern part of the Gulf of Finland (in Russian – „Finskij zaliv"), with Primorsk Oil Port, and the Baltic pipeline system (A). The dotted line on the map shows where the laying of pipeline to Finland will possibly be continued. New and old ports are marked by anchors, numbers indicate strongly protected nature areas, and fish silhouettes mark the spawning places. The areas (B) are coastal protected areas: 3– „Berezovye Islands", 9 – „Lebyazhye", 12 – „Kurgalsky". Sign (C) indicate the location of nuclear power plants.

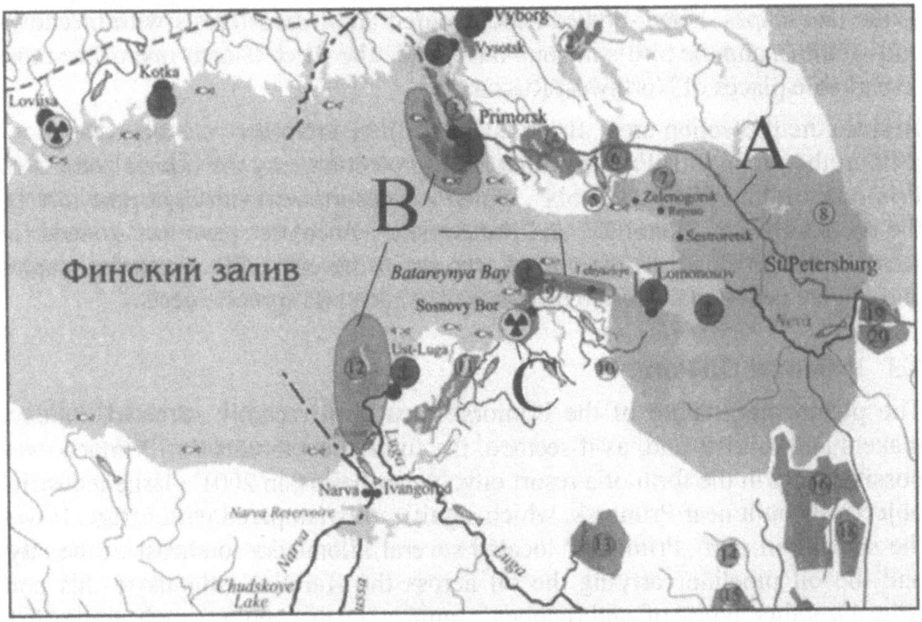

The state-of-the-art Primorsk oil complex is equipped with a fleet for eliminating oil spills and is connected to the Baltic transport system. It started its operation at the end of 2001. Since then it has been operating at maximum annual capacity of 12 mt. The ships docking at Primorsk can be loaded up to a maximum draft of 15-15.5 meters, which in terms of the load is equivalent to approximately 130000-150000 tons of oil (Photo 2).

Today the terminal has 18 reservoirs of 50000 cubic meters each. The Baltic transport system includes almost 800 kilometres of oil pipeline, which passes under the Neva near Shlisselburg. From the ecological point of view, this tunnel beneath the river is a point of highest risk in the oil pipeline system, though the applied technology, a combination of concrete tunnel, plastic tubes, steel tubes, oil pipes, appears functional and reliable.

According to the Government decision the capacity of the Primorsk oil terminal is now augmented up to 60 mt annually, and the potential limit of the overall capacity of the special oil port „Primorsk“ is in the range of 75-100 mt of crude oil and oil products annually. This will require not only oil transportation facilities, but also the development of oil processing facilities.

Photo 2: The Oil Terminal of the Primorsk Harbour

Laila Kule, September 2005

The majority of the developed projects for the Primorsk area include the development of a port centre around Primorsk as such. Also, there are plans for constructing a coal terminal to the east of the existing oil terminal. Furthermore, to the west of the oil terminal in the direction of Primorsk a recreation complex is being planned.

3. Description of the problem

The completed case study revealed two aspects of „development problems": ecological and social, which are closely connected.

3.1. Ecological aspects

The coastal ecosystem, in which the establishment of the second oil terminal is planned, has not been studied well by experts yet, due to the fact that it has been a „near- border region" closed to visitors since World War II. A special investigation of the biological communities was organised at the start of the first terminal construction, in 1998 – 1999 (Belenko et al., 2002; Lukyanov et al., 2002; Shilin, Zazerina, 2004). The results showed that the key biological species in the area are very sensitive to anthropogenic impacts.

Spawning *smelt*, concentrating during the spring for breeding near the construction site, attracts thousands of fishermen.

Sea ducks and geese spend so much energy during the migration that a peaceful rest is absolutely necessary. If the birds are disturbed, the whole migration can be endangered: they simply will not arrive at the place of nesting.

Ringed seals can be seen around the „Berezovye Islands" quite frequently. At the end of winter, the near-coast ice fields to the south and west of the Islands serve as a breeding place for the remaining ringed seal population in the Russian part of the Finnish Gulf (approximately 300 animals at present). Intensive ice mobility caused by the operation of ice-breakers may harm baby animals, if the new fairway is located too close to resting areas of baby seals?

Sprat, or *Baltic herring*, spawns in the Bjorkesund strait, around the „Berezovye Islands" and on the shallows along the coast close to the outlet to the sea (3-15 m), in general in the middle of June. Deepening and dumping, extraction of iron and manganese concretions and other factors, which cause mobility of the ground, can damage traditional places of spawning.

Coastal meadows- the most aesthetically attractive (and sensitive at the same time) coastal biotope, with a high degree of biological diversity rare species of plants, which can be found in the Red Book, grow here, including the sea violet *Viola maritima*. The coastal meadows suffer from a number of impacts caused by holiday-makers.

Benthic organisms, small *Crustaceans* and *Mollusk*, are especially sensitive to turbidity, ground mobility and oil pollution. Small (6-9mm) light-brown *Pontoporeia*, referring to *Amphipoda*, is an important food component for benthotrophic fish. When dumping takes place, and ground suspension rises causing increased turbidity, even the smallest trace of pollution makes *Pontoporeia* change its habitat.

Planktonic Crustaceans, *Daphnia* and *Bosmina* suffer when dirty waters from small private motorboats enters the coastal water. As a result, the food supply of planktotrophic fish is negatively affected.

Thus, the ecosystems of the case area are under constant and various anthropogenic pressures. They are influenced by different stress factors, which should be controlled (Viktorova, 2005; Shilin, Shilina, 2006).

3.2 Communication and social aspects

It is evident that the construction of the new ports on the shores of the Gulf of Finland will cause some negative impact for a number of vulnerable natural complexes, species of plants and animals. This may cause conflicts between different stakeholders. These problems have to be solved in interaction between the general public and local authorities. To make conflict resolution activities successful it is necessary to develop a continuous public participation process. The main stakeholders in the studied area are:

- Primorsk oil port,
- local authorities,
- local population,
- environmental protection organisations,

and other organisations, such as:
- Technical Centre of Space-rocket Corporation „Energy“
- Scientific Testing Area of the Ministry of Defense of the Russian Federation
- „Petrotral“ fishery company
- „Primorsk Fisherman „ fishery company
- „Primorsk Fish Cannery“ company
- Forestry of Primorsk
- Coastal Reservation „Birch Islands“.

The transformation of Primorsk into a city port brings some changes in the population's priorities and lifestyle, which are more dynamic in city ports. It is evident that the population must participate in such an adaptation to the new conditions. For this purpose it is not sufficient to implement information and PR programmes explaining the influence of port complexes on the living conditions. It is necessary to develop programmes of public participation, to create public committees so that they can participate in decision making, representing and taking into account the dynamics of public opinion.

According to the federal law „On ecological assessment", the public ecological assessment is an obligatory part of the state ecological assessment in the process of design and construction of objects. A sine qua non condition for the realisation of the public ecological assessment are public hearings with participation of experts, public organisations and the general public, as well as representatives of the local administration. The preparatory stage of the realisation of the public assessment assumes that the population should be informed about the construction plan of the object, its location and operation conditions. Also, additional information, which may be of interest to the population and society, should be included. Overall, in the case of construction of objects in Primorsk petroleum port these requirements have been observed.

As part of the ecological assessment related to design and construction of these port facilities, a survey was organised for the purpose of finding out the impact produced by the construction of these objects on the population. Such sociological surveys were started in 1999 at the design stage of the Oil port, and have been conducted since then. The questionnaires include questions concerning the general development of the coastal area of Primorsk, as well as questions aimed at discovering attitudes towards various construction objects (the first section of the oil terminal, the Baltic pipeline system, the factory for light petroleum processing, etc.).

4. Relevant conflict resolution methods and tools

The objective of the ecological component of research was to assess the environmental sensitivity of the coastal zone. The information obtained on the basis of experimental data is presented in the form of environmental sensitivity maps of the coastal area where the main types of human impact are described. The analysed study results are available to: (1) the local authorities to be used in the decision-making process; (2) the local population to be informed about the environment and to make the results of subsequent public opinion polls more valuable.

The sampling of the complex ecological information was carried out in 2005 – 2006 once per each season (spring, summer, autumn, winter), at 18 stations along 5 transects. From spring to autumn, this material was collected from a small specially equipped vessel. The information about the abiotic components of the ecosystem (structure of water masses, concentration of oxygen and biogenes, pH) was collected using STD- probes SeaBird and YSI (USA), „Seamoon" (Island) and Doppler prophilographe of water velocity. In the winter the investigation was made from the ice surface, by boring through the ice cover. Special attention was paid to biotic components of the ecosystem. It was necessary to choose the group of organisms, which would be most convenient, sensitive and suitable for bio-indication. It should be: (1) easily available for observations; (2) relatively stable in time, and connected with the studied biotope; (3) convenient for statistical operations. Phyto- and zooplankton are very variable in time and space, so information about changes in its structure is not indicative. Coastal vegetation (macrophytes) is not an obligatory component of the coastal ecosystems, especially in areas with a narrow littoral zone. Fish are not really indicative because of intensive migrations and concentration in different parts of the area in different seasons. In fact, only fish nurseries and their general conditions are indicative for the coastal zone. Birds are indicative in cases of a ship accident, where dramatic oil pollution can take place, but not in routine observations. The number of birds depends not only on the local conditions, but on events in other areas of the species' life (especially – during migrations). Marine mammals (seals) have too complicated behaviour; they can avoid „unpleasant" places, so the presence \ absence of seals does not show the real trend in the ecosystem's condition. The most informative component of coastal biota is the benthos, which is stable in time, reflects the local situation of the ecosystem, and can be used for analysis of the ecosystem changes in retrospective. The condition of benthic communities was analysed before the harbour construction, during the construction, and after. Seven types of bottom communities were mapped. As an ideal bio-indicator, using the experience of Swedish experts, the small Amphipoda *P. affinis* was chosen. *Pontoporeia* is an important part of fish food in the studied area, so its absence affects fish productivity. If anthropogenic pressure is strong, the population of *Pontoporeia* decreases dramatically in a very short time, which clearly indicates the negative changes in the ecological conditions.

The objective of the sociological component of research was implementation of methods of Integrated Coastal Zone Management, which requires the development of direct cooperation between local authorities and the general public on a local level (Viktorova 2007; Plink & Viktorova 2007). Taking into account the public opinion in the process of strategic planning of the development of new port facilities will contribute to the development of the public participation in the decision-making process. The study of public opinion is a necessary step in this direction.

It is also important to estimate the level of sustainability of the public participation process. It is known that at least three levels of public participation can be defined (non-participation, „talkinism" and active participation through partial delegation of powers). The scheme of the development of sustainability of the public participation process within the scope of the ICZM methodology is shown in Figure 2.

Fig. 2: Development of sustainability of public participation process within the scope of the ICZM methodology

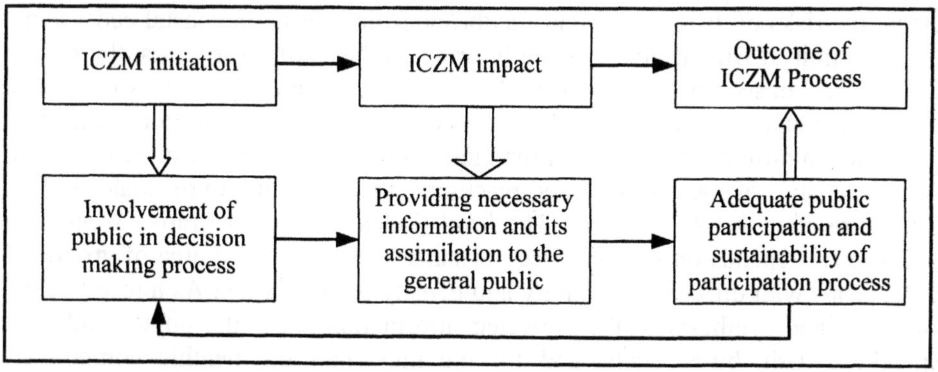

Tools used in the Primorsk case area for studying the public participation process can be defined as follows:

- Focused interviews
- Structured survey
- Study of documents, mass-media
- Analytical work.

The methods of interviewing local people are well approved of and have already been used by the working team during similar studies in other case areas in the coastal zone (Averkiev et al. 2006). In the Primorsk area, interviews with the locals took place mainly in summer, using the method of „accidental choice of respondents", or „participating in the situation".

Analytical work includes the elaboration of strategy and design of questionnaires, analysis of field data obtained, as well as elaboration of recommendations on improving communications and strengthening interaction between the local municipality and the residents of Primorsk.

The sociological surveys carried out in Primorsk and its environs from 1999 until 2006 were aimed at the overall objective to discover the population's opinion concerning the construction plans, operation and expansion of the oil port „Primorsk". However, in 2006 the questionnaire was completely revised and reoriented, except for the most interesting questions, which have been used since the first survey in 1999, regarding the identification of the public attitude towards the basic socio economic cultural objects and directions of development of Primorsk municipality.

5. Results

5.1 Results of Ecological Studies

The ecological sensitivity maps of the coastal zone ecosystems are constructed in GIS format for planning safe and sustainable exploitation of the Primorsk oil terminal, and supporting the decision making process. These maps help to estimate the risks of different kinds of anthropogenic effects on coastal ecosystems in the Strait of Bjerkesund. For the information basis of these maps, the results of the field observations were used. We observed the ecological sensitivity as a possible integrative reaction of different components of coastal biological communities to different types of anthropogenic pressure. One of the most important characteristics of such ecosystems' reaction is the possibility to re-establish their structure after the stress effect. The level (index) of the integrative ecological sensitivity depends on the presence / absence in the observed area of groups of organisms with different sensitivity and density of population. As a result of the analysis and synthesis of the collected information about the phyto- and zooplankton, fish, birds, seals, and for the most part the benthic community, schemes for the distribution of these components of the coastal biota in different seasons were constructed. These schemes were incorporated in GIS as different „layers" on the regular net. The index of the integrative sensitivity depends on the number of objects in every section of the regular net.

Zones of different attention for the decision makers are mapped for the Bjerkesund area over the four seasons for four potential types of anthropogenic pressure (oil film, oil emulsion, suspended matter and dumping) – normal attention, concentrated attention, and special attention. These zones are marked with different colours (red, orange, and green).

The most sensitive areas of the Strait of Bjerkesund are the southern „gates" and the north-eastern coast, because of the high concentration of phyto- and zooplankton and benthos in the summer time. In benthic communities, the presence of small *Crustacea* and *Bivalvia* was observed. In these areas, the fish nurseries are located as well. In the spring time, migrating birds are concentrating here for relaxation and feeding in the shallow water. Low sensitivity is typical of the central (deep) part of the strait, where no benthic communities are registered. Seals can be observed only during the winter time, when the integrative sensitivity of the studied ecosystem is low. The integrated sensitivity of the area during different seasons is decreasing in the order: summer > spring > autumn > winter.

The structure and the functioning principle of radar ecological monitoring of coastal zones for the control of ecological situations were proposed (Beskid 2005; Viktorova 2007). The local permanent control was conducted in June-October 2006 by means of a stationary radar post with the streamlined C-band radar station „Furuno FR-7062" and the measurement device including visual convoy of the observed object. The coastal radar had been functionally modernised for the marine surface observation task and for pollution anomalies detection. During all observation periods, permanent (twenty-four-hour) radar control of the task zone was conducted, which was provided by 2-3 operator shifts with 8-12 hours in each shift respectively. For the purpose of radar image acquisition of signals reflected from pure and oil –polluted marine surface, special software was developed. Ships that passed through the control zone and other oil polluting sources were observed. During the observation time no significant oil pollution was detected. However, small rainbow coloured spills from unknown sources were observed.

5.1.1 Lessons learned

1. At present, the main ecological problems in the case area are caused not only by the operating oil terminal, but also by poor ecological knowledge and culture of the locals, who are mostly newcomers. The surface waters in the area are polluted by oil products coming from small private boats.
2. In the future, the construction of new terminals in the Oil Port will increase anthropogenic pressure on the coastal ecosystems, with a negative impact for a number of vulnerable natural complexes, species of plants and animals. Other types of anthropogenic impact in the case area, such as the fish processing factory (dumping untreated sewage), sewage disposal installations, pollution of the water conservation zone with household sewage and sewage from the railway station, as well as the felling of forests combined cannot cause the ecological damage comparable to the potential risk of an oil overload and transportation, dramatically increasing with further development of the oil port complex.
3. The operators of Primorsk Oil Port, local inhabitants who pick mushrooms and berries, tourists and holiday-makers who come to the beaches and forests, managers planning the fairways, owners of motor boats – everybody should be cautious and avoid disturbing those places with high ecological sensitivity/ vulnerability. In case the ecological potential of the coastal zone is not undermined, its ecosystems will retain high biological productivity, aesthetic attraction, and the „ecological service" will still be unique and valuable.

5.2 Results of sociological research

The most active participants of this survey were native inhabitants of Primorsk (almost half of the respondents). About one fourth of the questioned people were not born in the town. The majority of respondents are between 20 and 60 years old. Most have a second level education, and approx. a half have third level / university education (both completed and uncompleted). More than half of respondents are employed, only 12% are unemployed. About 70% of the respondents are officials.

The respondents have been asked about their family status in order to define the social group, which would be mostly interested in sustainability and safety of their region, because those respondents who have a family care also about their relatives.

5.2.1 The population's attitude towards the Primorsk municipality services and social activities

Two alternative questionnaires used for the survey in 2006 were in the form of questionnaire- interview in order to provide for a more detailed study and analysis of the social situation. The concise form aimed at carrying out a brief survey of the socioeconomic and cultural situation within the municipality. Due to using the questionnaires it was possible to develop a „typical person" reflecting public attitudes towards the development, as well as public anxiety and concerns regarding their residential areas.

As one of the survey's results the rating values for different municipal services and activities have been obtained (Table 1).

Tab. 1: Rating of various services and social activities of the Primorsk municipality (1 = excellent, 4 = unsatisfactory)

Service / Activity	**Rating**
Preschool and school education	**1.84**
Fire and rescue service	**1.96**
Primorsk oil port complex	**2.54**
Transport infrastructure	**2.69**
Work of the Municipal Administration with the population	**2.85**
Influence of the Municipal Administration on society	**2.86**
Environmental protection	**2.96**
Communal service	**2.97**
Medical service	**3.10**
Cultural activities	**3.23**
Social service	**3.43**

Overall, according to the survey results it is possible to state that the most positive rating has been granted to the work of the educational sector and the fire and rescue service. In the opinion of the local population the work in these areas (using the school scale) can be estimated somewhere between good and excellent. The most problematic situation (in the opinion of the survey participants) is in the field of social services (close to unsatisfactory).

The complete data received during this sociological research will be used in the work of the local administration. The conclusions made on the basis of the analysis of the public survey results will be taken into account not only in the daily work of the local administration aimed at normal operation of all the municipal services, but will also help to make some corrections to the work aimed at improvement of interaction between the administration and the public.

Within the scope of this case study it is not possible to discuss all survey data related to the consultation reflecting the local aspects of this specific municipality. Distribution of answers to some questions, their analysis and some conclusions are presented below.

Cooperation between the local administration and the population will be more stable if the evolving dialogue addresses the most acute (in the population's opinion) problems. Figure 3 shows the distribution of the population's opinion concerning the work of the community services (waste removal, water supply, sanitation). As we can see, more than half of the interviewed inhabitants estimate this work as unsatisfactory; this confirms that this problem exists. The number of participants without a definite opinion is very low – only 4%. In other examples (which will be shown further on) the percentage of people without a definite opinion reaches 20%, which rather often exceeds the percentage of people having quite a determined (positive or negative) opinion on the problem.

Fig. 3: Distribution of answers to the question „Operation of water service facilities, sanitation and street cleaning is good and reliable"

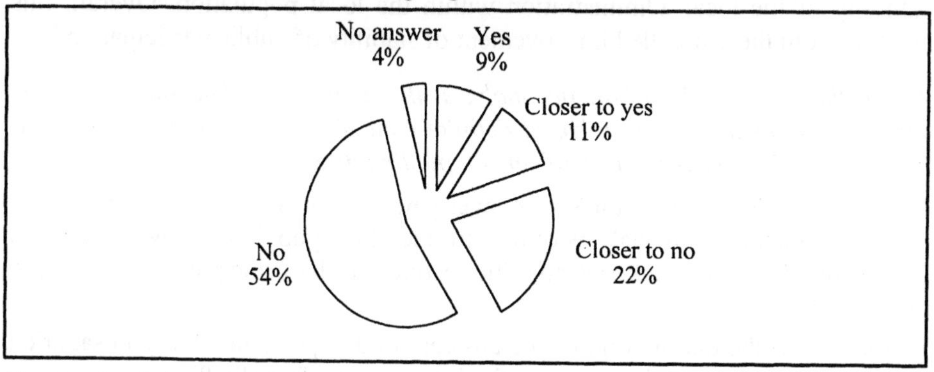

The depth of the problem can be evaluated with the help of the data obtained from the interviews. The results of the analysis of these data are shown in the Fig. 4.

Fig. 4: Distribution of answers (%) to the group of questions „Quality of cleaning, water supply, sewerage"

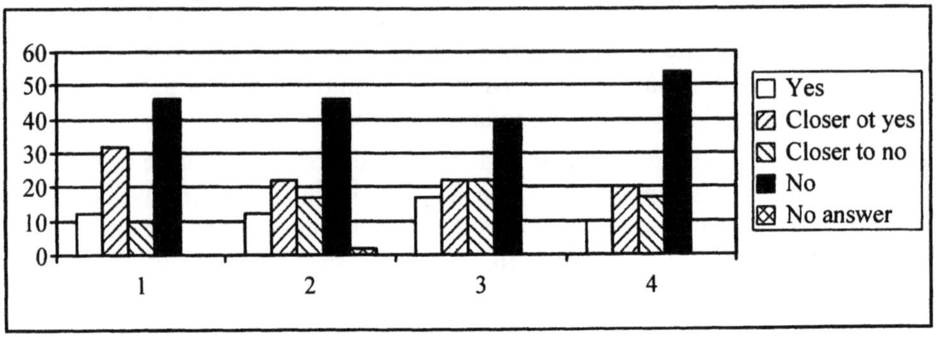

1. *Quality of street cleaning is good*;
2. *Water supply and sewerage systems are good and reliable*;
3. *Waste is removed regularly and does not pile up*;
4. *In general you are satisfied with the quality of cleaning, water supply and sewerage systems in the municipality.*

The results show that the general estimate of the quality of the community services both in the surveys and interviews is rather low. It is evident that the problem is rather deeply rooted, as the public opinion about operation of all the community services is rather unanimous. It is necessary to note that in the results of the interviews there are practically no uncertain answers. The specialists, who carried out the surveys and interviews, have stated in their reports that the citizens are highly interested and ready to discuss this problem. Without doubt, the introduction of questions relating to the work of community services into public discussion will be interesting for the population. It is expected that municipality residents will willingly participate in the discussion if the subject is acute and related to their interests. On the other hand, the real improvement of community services (also as a result of taking into account the public opinion) will foster an increase in trust and authority of the local administration within the local population. Overall, this must result in the „so-called improvement of stability of public participation".

Within the framework of holistic analysis, the analysis of data obtained from various sources allows to arrive at definite conclusions, and in some cases, to find out the most general cause of certain public attitudes.

The following is two examples of convergence of various methods being used for data acquisition and analysis of the attitude of the population towards various aspects of life, which makes it possible to find out the principal cause of negative attitudes.

1. Analysis of the distribution of the answers to the question „Are you satisfied with the level of social services in the Primorsk municipality?"

Figure 5 shows the distribution of opinions regarding the level of social services provided to the citizens of Primorsk according to the survey's results. As it has already been stated (see Table 1), this area of work has the worst rating in comparison with the other services. The results of the interviews (see Fig. 5), in principle, lead to a rather obvious conclusion regarding the poor operation of the social services.

Fig. 5: Distribution of answers to the question „Are you satisfied with the level of social services in the Primorsk municipality?"

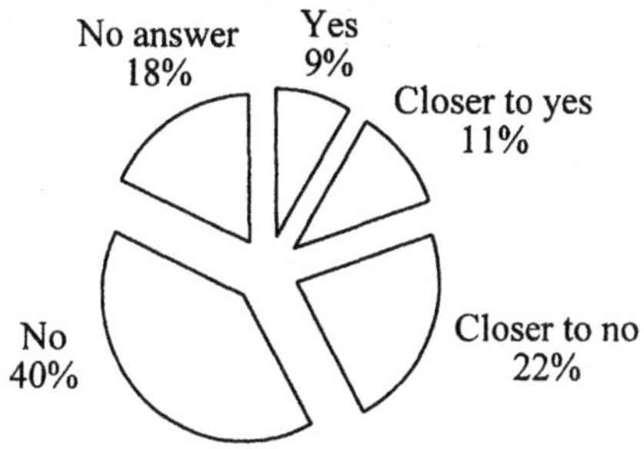

This conclusion may not be fully correct, so we shall pay attention to the additional results obtained in the course of the interviews, which are shown in Figure 6. 70% of the respondents answered „No" to the question „Do you often use the social services?"

However, this cannot be taken as a negative sign, because the majority of the survey participants were young people below 30 (about 30% of the interviewed people), and middle-aged people (30-40 years) made up almost a quarter of the total respondents. The fifth section of the respondents comprised visitors or those living in this municipality for less than one year. Many parameters of social services (dimension of allowances, pensions and other) are prescribed on a federal level. This is why such a ubiquitous dissatisfaction of the Primorsk population about the social services is caused by problems that are outside the authority of the municipal administration. The specific attitude of the population towards local municipal services is reflected only in answers to the second question, which was used to identify opinions of those people who really approached these services and had definite attitudes towards their work. If we do not take into account the opinions of 30% of the respondents who gave no answer (those who gave a negative answer to the first question) then the situation will still look complicated, but not as grave as it seemed at first sight.

Fig. 6: Distribution of answers (%) to the group of questions „Quality of social services in the municipality"

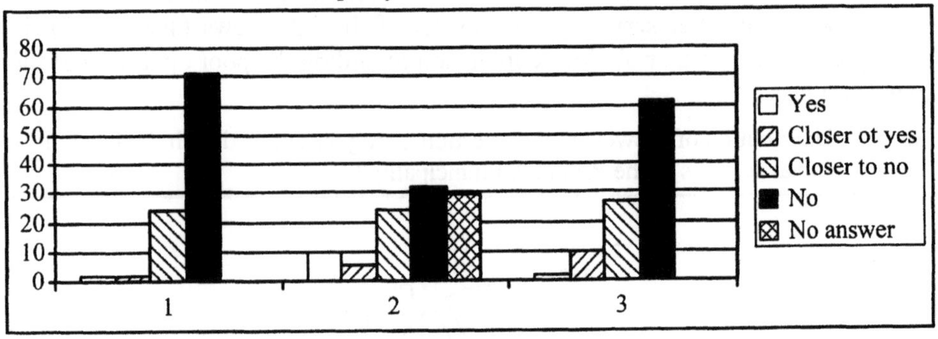

1. *You often use social services;*
2. *When you approach a social service you always receive help;*
3. *In general you are satisfied with the level of social services and protection in the municipality.*

In Fig. 7 the distribution of answers to the question concerning public transport can be seen. As the diagram shows, the majority of the population is not satisfied with the public transport. Using the interview data (6) we can conclude that the principal cause of this dissatisfaction is rooted in the poor quality of motor-roads. We have to note that the distribution of the interview answers to the last question of this section (practically repeating the question from the survey questionnaire) matches the distribution of the survey answers well. This confirms that the internal structure of answer distribution is similar and that it is possible to combine interview and survey data within the framework of holistic analysis.

Fig. 7: Distribution of answers to the question „Are you satisfied with the level of public transport and the quality of the roads in the Primorsk municipality?"

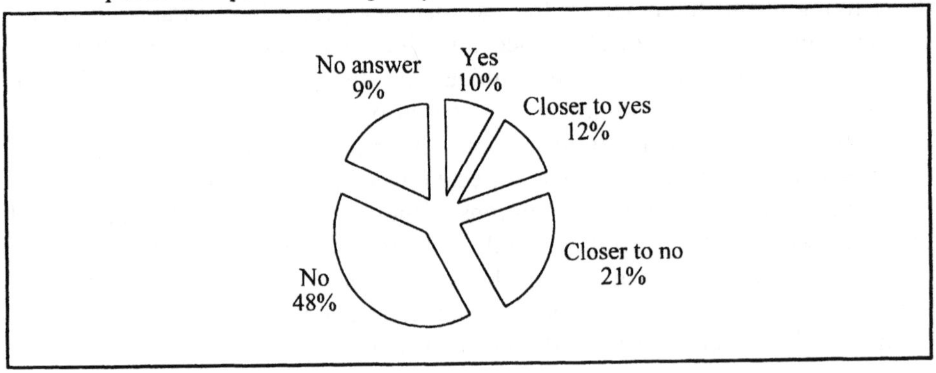

Fig. 8: Distribution of answers (%) to the group of questions „Quality of municipal transport infrastructure".

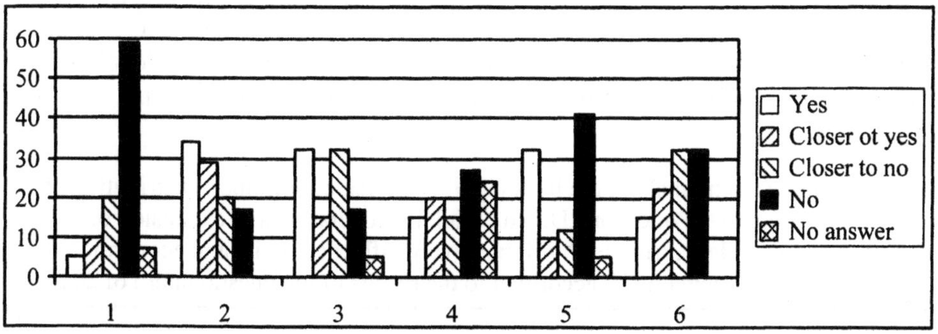

1. *The roads in the municipality are maintained in good condition;*
2. *As a pedestrian you feel safe on the roads;*
3. *You can easily travel on business to another municipality using public transport;*
4. *You are satisfied with the level of train service;*
5. *You often go to Saint Petersburg using public transport;*
6. *You are generally pleased with the work of the transport system in the Primorsk municipality.*

5.2.2 The population's attitude towards the oil port development

Other points of this questionnaire are intended to learn the opinion of the public of Primorsk about the building and future use of Primorsk oil terminals, as well as the amount of information provided to the public by the Mass Media about this Port complex. Talking about the spreading of news, only 11% of respondents are satisfied with the information that they receive through the Mass Media (Fig. 9).

Fig. 9: Distribution of answers (%) to the question „Information in the Mass Media: is it enough?"

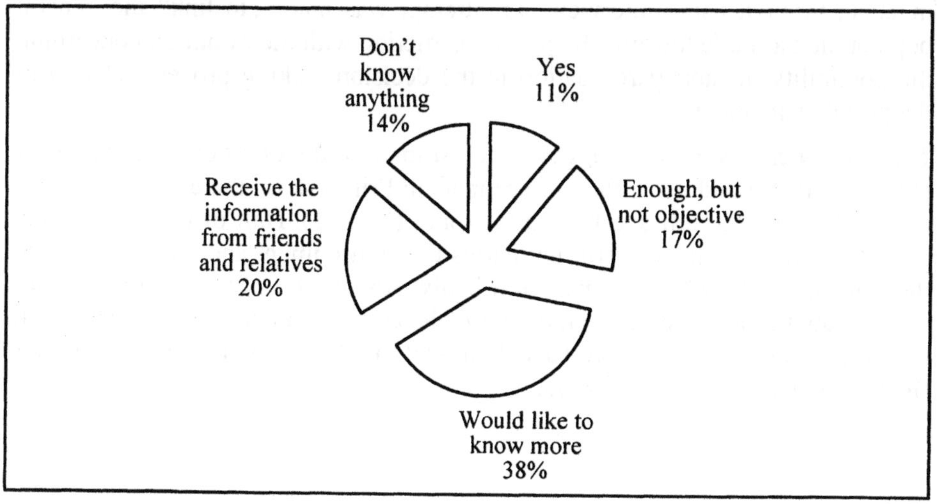

The common opinion of the municipal population on the subject of the oil complex development in the Primorsk port is shown in Figure 10. The Integrated Coastal Zone Management Department of RSHU has been carrying out such surveys since 1999. This period includes all the stages of the port development from planning and design up to commissioning of all the facilities that are operating now.

Fig. 10: Public Opinion about the building of Primorsk oil terminal. Distribution of answers to the question „Do you think that operation and expansion of the port complex „Primorsk“ will positively influence social and economic life of the municipality?", according to the results of the questionnaire of 2006.

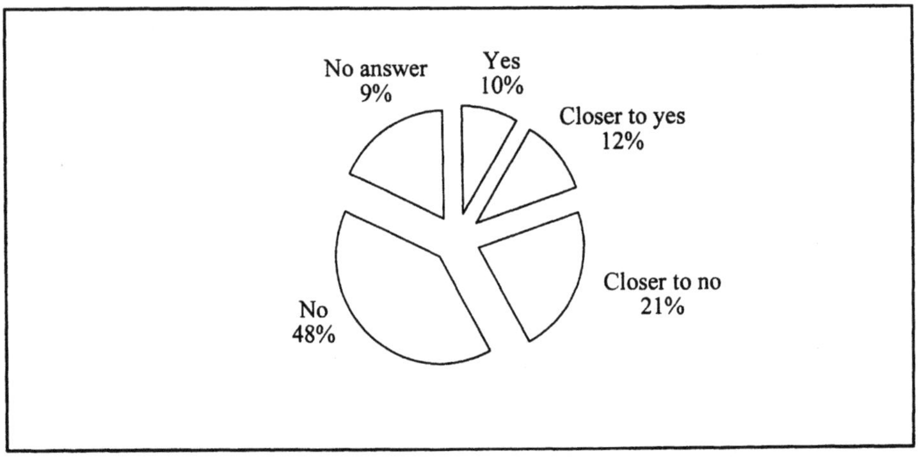

The results of the 2006 survey show that the general tendency towards a predominantly positive attitude regarding the new port development persists in approximately the same correlation. In contrast to the previous research however, in 2006, through using interviews an attempt was made „to link“ the general population's attitude towards the port construction with their opinion concerning the possibility of their participation in the decision-making process relating to the port development.

The respondents were also suggested to estimate in the questionnaire the influence of building and using the oil terminals in Primorsk on the ecological situation in the Finnish Gulf and its coastal zones (Fig. 11). Half of the respondents think that the Primorsk oil terminal harms the environment and would influence the situation in the whole Finnish Gulf badly. About 40% of the respondents are worried about the possible degradation of local environment near the Primorsk oil terminal in the future. Only 10% think that the Port complex will have no influence on the ecological situation.

Fig. 11: The influence of the port complex on the environment

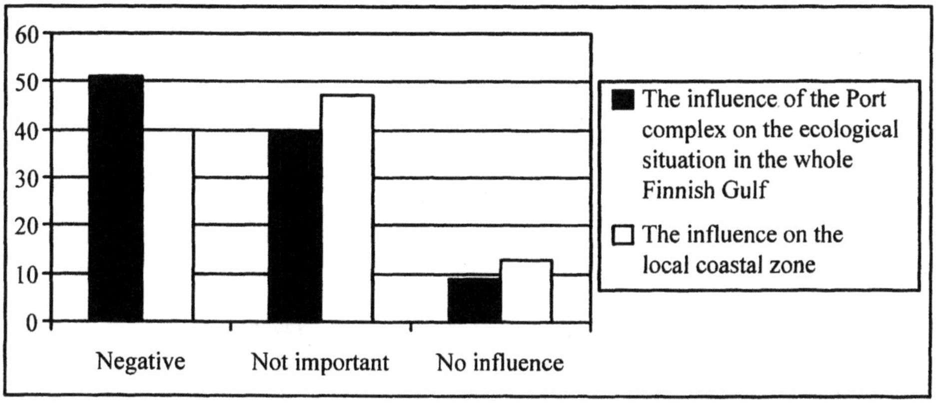

The remaining diagrams (Fig. 12, 13) illustrate the expectations of local people connected with the building of oil terminals in Primorsk. The majority of respondents expect that the building of the oil terminals will change the status of Primorsk, and in future it will develop into a modern industrial city with a much bigger population and more possibilities of finding a good job. Many inhabitants of the town need a job, but the port complex has created almost no new jobs.

Fig. 12: Will the Port complex provide new well-paid jobs?

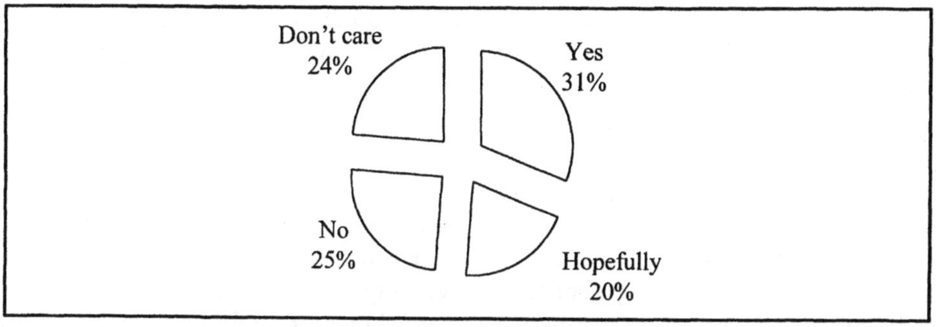

However, only 19% of the respondents think that the operation of the Primorsk oil terminals will bring real benefit (Fig. 13).

Those respondents who support the project of building the oil terminals in Primorsk are mostly young people who don't have a family yet and are less responsible for their decisions. Besides, most of such respondents are unemployed or still study and hope that transporting oil will turn Primorsk into a highly developed industrial city where they could find a good job.

The respondents who are against building the oil terminals in their town are mostly native inhabitants of Primorsk who have families and work mostly as fishermen, because fishing is a traditional occupation of people who live in this port town. They do not want any changes in their lives and do not see the new port complex as environmentally friendly, because the building of oil terminals has disturbed the fish around and interfered with fisheries.

Fig. 13: Expectations of economical profit

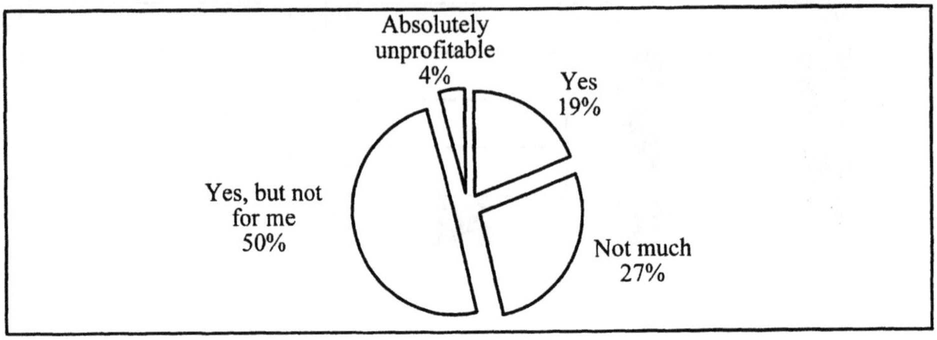

However, all the respondents want safety and sustainability for the local environment.

The results of the statistical analysis show that people in Primorsk are not aware of the consequences of the oil terminal's construction and problems of sustainable oil transportation. The port complex of Primorsk and the local population do not have any contact, any common ideas or interests. Only 19% of the respondents expect the new oil project to be economically profitable for the population. Only 11% receive enough information about the building of the oil terminals. 40% say that they „don't know" the real situation and don't really care about it. However, the question of oil transportation is one of the most important questions nowadays. The more people are involved in this problem and in such projects as in Primorsk, the less risk there is to damage public interests and environmental sustainability. That is why it is necessary to raise public awareness of possible consequences of oil terminal development in Primorsk. The port complex should make the public interested in the problems of sustainable oil transportation by raising employment level, involving local people in the process of building and using the Port complex, and spreading more information about it in the mass media.

5.2.3 Assessment of sustainability of the public participation process

The most approved of and widely used procedure, which ensures taking into account public opinion, is the procedure of public ecological impact assessment (PEIA). The assessment of adequacy of the PEIA procedure in Russian practice is not developed enough so far. This fact brings about a large number of problems when the procedure effectiveness is being assessed.

The opinion distribution obtained through the interviews is shown in Fig. 14. As can be seen from the legend, the questions in the interview were divided into two groups: the first three questions (1-3) were related to the assessment of population participation in PEIA, the last three questions (4-6) touched on population attitude towards the port development. Brief conclusions on the basis of these groups of questions can be formulated as follows.

Fig. 14: Distribution of answers (%) for the group of questions „Operation of the port complex 'Primorsk'"

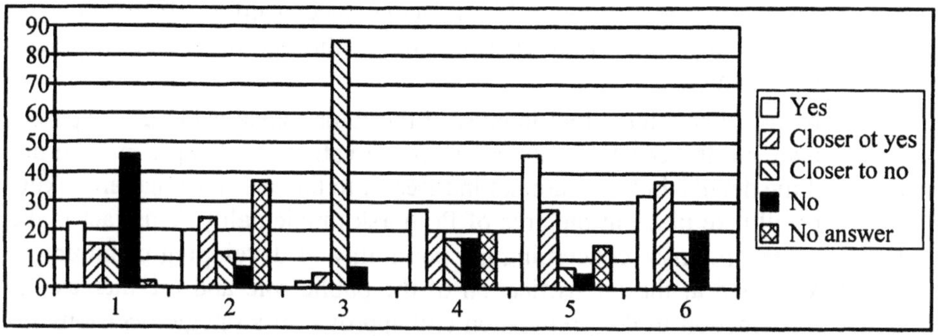

1. *Do you know about the „public assessment" concept?*
2. *Do you have a positive attitude to the public assessment of the construction of fundamental industrial objects in the Primorsk municipality?*
3. *You took part in public hearings and assessment regarding the construction and expansion of the port complex „Primorsk";*
4. *You think that the operation and expansion of the port complex „Primorsk" influences social and economic life of the municipality in a positive way;*
5. *Your attitude towards the port complex „Primorsk" is positive;*
6. *In general you are pleased with the fact that the port complex „Primorsk" operates and develops in the municipal area.*

Approximately half of the respondents have a poor comprehension of what PEIA is (see answers to Question 1). Therefore, about the same percentage of interviewed participants have difficulties in saying whether it is good or bad (see answers to Question 2). This is why the natural reaction to the question „Have you ever participated in a public ecological assessment procedure" for more than 80% of participants was „Closer to no".

In other words, the interviewed participants vaguely remembered that something had been taking place, but most likely they had not been involved in this process. Thus, the interview results have shown that the PEIA procedure has not become a basis for the development of communicative policy in the field of environmental safety and is used now very inefficiently. The reason is the absence of the approved standards for the PEIA procedure requirements concerning a minimal set of activities to be performed within the framework of PEIA.

5.2.4 Lessons learned

1. The knowledge of the local population about the plans, prospects and results of the Primorsk Oil Port operation is rather poor. This is why it is sometimes difficult for the respondents to give a definite answer to a specific (special) question. It is necessary to expand more information at the local level about the port activity at present and further plans for its development. Such information can be provided through Mass Media, special informative brochures devoted to the problems of the coastal zone, or through organizing a special Information Center.
2. Primorsk inhabitants have real interest in the future sustainable development of the town and the coastal zone, nature conservation etc. People may be unaware of the details, but they are not indifferent to the activities that are taking place or will take place in the area of Primorsk municipality. This fact is the basis for our optimism concerning the possibility of developing communication links aimed at the intensification of the public role and the increase of public participation in the preparation and approval of strategic decisions influencing the development of Primorsk municipality.
3. The experience of cooperation with local administration and the analysis of the results of sociological surveys show that there is a necessity for more precise regulation regarding the implementation of some of the procedures of public ecological assessment (volume of information and deadlines for its publication, criteria of representativeness and legitimacy of public opinion, clarification about how opinions should be taken into account and approval procedures, etc.). The approval of the Instruction for Public Assessment could be a useful tool for establishing a dialogue between the general public, conservationists and governmental bodies.
4. Analysis of the results of public opinion survey can help the local administration in developing its communication policy.
5. Non-professional public opinion surveys may result in incorrect conclusions due to various reasons: (1) unclear questions in the questionnaire; (2) different understanding of questions by the respondents; (3) influence of subjective factors reflecting the respondent's attitude to the problem.

6. Conclusions and recommendations

The planned oil port expansion and construction of the oil refinery in Primorsk, as well as construction of the second part of the Baltic Pipeline System BTS-2, will entail a considerable increase of anthropogenic impact on the ecosystem of the Bjorkesund Strait in the near future. Obviously, it is not possible to neutralize the impact caused by the growing oil port and other industrial objects completely. However, there is a possibility to reduce the risk of emergencies, including the overall impact of anthropogenic activities on the natural ecosystems by means of an integrated coastal zone management approach, based on scientific research. In practice, the results of the case study, including opinion surveys and environmental research of the ecosystem, used for the creation of sensitivity maps, will be used for elaboration of recommendations to the local authorities aimed at improving the decision-making process in the field of coastal zone management. This will contribute to:

- Improvement of strategic planning for the region;
- Procedures of Public hearing and Environmental Impact Assessment;
- Development of a municipal communication policy;
- Exposure of „hot-points" in interaction between municipality and residents;
- Improving public safety in the case area.

As a further result, the case study will contribute to the sustainable development of the Primorsk area through broad involvement of the general public in decision making and by increasing the sustainability of this participation process. Sustainable development of the coastal zone of the Bjorkesund Strait is only possible if we strike a balance between the process of developing the new port complex and conserving the coastal ecosystems with their rare and key species. Ecological and cultural diversity of the coasts of the Bjorkesund Strait must be safe, and this is the task of locals, scientists and experts, managers, owners of private boats, and tourists.

On the basis of the completed case study, the following steps can be suggested towards increasing control of the ecological situation of the coastal zone: Ban or restriction on deforestation, bottom deepening, quarrying, polluting water areas, dumping oil products, commercial fishery, using small vessels during the period from ice break up to June10, visiting nesting sites of birds from April 15 till August 15, picking mushrooms, berries and medical plants in the coastal zone, as well as allocating land for construction work and other forms of economic activity, which might damage natural ecosystems. Shallow coastal zones, bird nesting colonies, seal breeding sites, fish spawning places, areas of sandy and meadow type littoral should be given the status of specially protected objects in the case area. To improve the ecological competence and culture of the population, it is necessary:

- To strictly obey the rules of non-disturbance of the protected area „Birch islands";
- not to allow anthropogenic interference on the western coast of the islands;
- to allow ground works, dumping, iron and manganese concretions extraction only at a certain distance from fish spawning areas;
- to organise special parking places for fishermen;
- to call on the local population not to arrange picnics and sport events on the meadows, to avoid trampling and leaving rubbish in the coastal lowlands;

- To establish rules for correct use of boat moorings, to require that boat users control the condition and cleanliness of boats and launches at the moorings before letting them out into the strait.

7. References

Averkiev, A.S., Gogoberidze, G.G., Kluikov, Ye. Yu., Kononenko, M.R., Lukyanov, S.V., Osnitsky, L.A., Plink, N.L., Shilin, M.B. & Kelasjev, O.V. (2006). Exit from the Labyrinth // Russian editor M. Shilin, English editors Gillian Cambers and Claire Blackburn / UNESCO Coastal Region and Small Island Papers 21, 75 pp.

Belenko, S.L., Bychenkov, D.Ye., Lange, Ye.K., Lukyanov, S.V., Nekrasov, A.V. & Shilin, M.B. (2002). Local Ecological Monitoring of the Harbour Construction Areas in the Gulf of Finland / Proc. of the XII Int. Conf. in Fishery and Oceanology, AtlantNIRO, Kaliningrad. p. 29-30.

Beskid, P.P. (ed.) (2007). Ecological Researches. Part 2. Remote monitoring of oil spills on the water surface in Primorsk area. – St Pb: The Russian Universities Publishing and Polygraphic Ass., 44 p. (in Russian, with English summary).

Gogoberidze, G.G., Kononenko, M.R. & Provotorov, P.P. (2005). Study of the Local Population Participation in the Process of Decision – Making Concerning Construction of the Oil Port Complex „Primorsk" by Means of Sociological Polls // Proc. of the Seminar „Russian and EU Experience in the Field of Integrated Coastal Zone Management", held on September 26-27, 2005 in Primorsk and St. Petersburg: The Russian Universities Publishing and Polygraphic Association. – p. 49-56.

Kosheleva, A.G. (2005). City of Primorsk and Special Port „Primorsk": Problems and Prospects for Development // Proc. of the Seminar „Russian and EU Experience in the Field of Integrated Coastal Zone Management", held on September 26-27, 2005 in Primorsk and St. Petersburg: The Russian Universities Publishing and Polygraphic Association, p. 42-48.

Lukyanov, S.V., Shilin, M.B., Belenko, S.L., Ivanov, V.V., Mandryka, O.N. & Pnyushkov, A.V. (2002). Impact Ecological Monitoring of the Sites of the New Harbours Construction // UNESCO Baltic Floating University Bulletin N 4-5, p. 109-119.

Plink, N.L., Shkoldin, V.G., Skorodumov, A.N., Kosheleva, A.G., Gogoberidze, G.G., Kononenko, M.R. & Provotorov, P.P. (2007). The role of communication conflicts in the system of Integrated Coastal Zone Management (on the Primorsk municipality taken as an example) // Editors N.L. Plink, E.V. Viktorova / St. Petersburg, 2007: 44 pp. (in Russian, with English summary).

Pogrebov, V.B. & Puzachenko, A.Yu (2001). Integral assessment of potential biota sensitivity to offshore oil-field development operations // *Coastline*, No. 1, p. 10-11.

Shilin, M.B. & Zazerina, Ye (2004). Control of the Ecological Situation at the Harbour Construction Site, and Sustainable Strategy Development for the Coastal Zone at the City of Primorsk (St. Petersburg Region) // Wise Coastal Practices for Sustainable Human Development / Riga-Roja: UNESCO Chair for Sustainable Coastal Development, p. 104-106.

Shilin, M.B. (2005). Ecological Vulnerability of Ecosystems of the Strait of Bjerkesund // Proc. of the Seminar „Russian and EU Experience in the Field of Integrated Coastal Zone Management", held on September 26-27, 2005 in Primorsk and St. Petersburg: The Russian Universities Publishing and Polygraphic Assopciation, p. 63-66.

Shilin, M.B. & Shilina, M.M. (2006). Studying the Impact of the Oil Terminals on the Coastal Ecosystems // Proc. VI Int. Youth Environmental Forum „Ecobalthica 2006", St. Petersburg, State Polytechnical University, p. 97.

Shilin, M.B., Kijko, O.A. & Lange, Ye.K. (2006). Plankton and benthos. Nature Conservation Atlas of the Russian Part of the Gulf of Finland / Ed. V. Pogrebov, R. Sagitov, N. Dmitriev/ St. Petersburg: Tuscarora, 48 p.

Viktorova, E.V. (2005). TACIS CBC Project „Decision Making Tools for Implementation of Integrated Coastal Zone Management (by the Example of the Primorsk Municipality)" Proc. of the Seminar „Russian and EU Experience in the Field of Integrated Coastal Zone Management", held on September 26-27, 2005 in Primorsk and St. Petersburg: The Russian Universities Publishing and Polygraphic Association. p. 9-14.

– (ed.) (2007). Ecological Researches. Part 1. Environmental sensitivity monitoring of the coastal zone in Primorsk area. St Pb: The Russian Universities Publishing and Polygraphic Ass., 84 p. (in Russian, with English summary). Materials of the Seminar „Coastal Zone Management for Sustainable Development", St. Petersburg, 1-2 March, 2007.

Coastal Zone Management in Ventspils

Ilga Zīlniece[1], Tatjana Valdmane[2], Ineta Kraule[3], Sofija Galindoma[4], Laila Kule[5]

Photo: A. Popeluha, 2006

1 Ventspils City Council, Head of Environmental Department, Ventspils, Latvia.
2 Ventspils City Council, Chief Specialist of Environmental Department.
3 Ventspils City Council, Specialist of Environmental Department.
4 Ventspils City Council, Deputy Head of Management and Information Department.
5 University of Latvia.

Abstract

The unique geographical location of Ventspils located on the Latvian eastern shores of the Baltic Sea has shaped its historical role as one of the most important export ports, from Russian Imperial times in the 18th century, through the Soviet Union regime until today. The transshipment of oil products and chemicals is and will be one of the most important aspects for the economy and the development of the port town. Therefore, the largest investments will mainly occur in the field of environmental protection in relation to the above mentioned transshipment.

Environmental problems of Ventspils are described and analysed in the Environmental Policy Plan (EPP) adopted by the municipality. The current Ventspils Environmental Policy Plan is the third one covering the period until 2010. The EPP combines a continuous process of monitoring, implementation and planning of new improvements. The plan contains a process of developing ways and means to improve the environment and diminish negative impacts through the involvement and co-operation of local government, enterprises, the Free Port Authority and citizens. The plan is based on the principle of sustainable development. One of the main goals of this plan is to develop a society with high environmental awareness.

The municipality is responsible for the implementation of local environmental policy and is obliged to devote increasing attention to the developments in the port, to ensure that there are no negative or destructive changes to the environment of the city and its surroundings.

1. Introduction

Ventspils is the sixth largest city in Latvia situated in the north-western part of Latvia on the Baltic Sea at the outlet of the River Venta. The total area of the city is 55.74 km^2, with 44.000 inhabitants in 2006. Ventspils is an important transport hub with an ice free deep-water port, which has one of the largest cargo turnovers in the Baltic region. At the same time Ventspils is a green city – almost half of the total area (45%) of Ventspils is covered by natural flora that consists mainly of pine forests (24% of nature area). The River Venta, with a length within the city borders of 3.75 km and a width between 150-300 m, divides the town into two parts. In the northern part of Ventspils Busnieku Lake is located with a total water area of 4.8 km^2.

Photo 1: Ventspils from birds' flight

Photo: A. Popeluha, 2006

The total length of coastline in Ventspils City is 11.85 km and is characterised by the beach and a dune zone along the shore reaching a maximum height of 15-20 m. The inhabitants of Ventspils have always been proud of their wide sandy, clean and orderly maintained beach. They use several bathing places within the area of the city: the River Venta and Busnieku Lake, as well as several places along the seashore. According to Regulations No. 300 of the Cabinet of Ministers of the Republic of Latvia issued on August 11, 1998 „Regulations of Arranging and Hygiene of Bathing Places" Ventspils City beach has been arranged and is being maintained for recreational needs. The beach used for recreation is a 300m wide and 1.2 km long section of the seashore within the area of

Ventspils city, situated to the south of the territory of the Free Port of Ventspils and near to where the River Venta flows into the Baltic Sea.

2. Background of the problem

The uniqueness of the Baltic Sea coast in Latvia and surroundings of Ventspils is determined by its valuable natural resources. The length of the Baltic Sea coast in the administrative territory of Ventspils is approx. 12 km, and almost a fourth of the city's administrative territory is occupied by Ventspils Free Port. When planning city development, the interests of the general public and entrepreneurs should be taken into account, at the same time ensuring the preservation of valuable natural resources and preventing and minimising environmental pollution. The development of economic activities in the Free Port territory together with the implementation of the policy of preserving valuable natural resources is possible only if the sustainable development principle – development that meets the needs of the present generation without compromising the needs of the future (Brundlandt report) – is put into action. It is the only way to successfully plan the development of the Baltic Sea coast and to preserve its valuable natural resources, particularly its coastal dune areas.

3. Description of the conflict

Ventspils historically is located on the coast of the Baltic Sea and on both banks of the River Venta. This case study developed within the scope of the Coastman project relates not only to the narrow part of the seashore referred to in the legislation of Latvia as the Baltic Sea protective belt, but assuming that the term „coast" here refers to the river shores as well, it takes into account the city as a whole. The case study describes methods and planning instruments to be used on the municipal level to fully prevent conflicts and to solve any potential problems before a conflict can arise. Therefore, the essence of the Ventspils case study comprises how to develop a town preserving valuable natural resources while at maximum level satisfying interests of all parties involved (different groups of population and institutions, including developers and environmentalists), as well as ensuring economic growth and meeting of social needs. The methods of conflict resolution were chosen with the aim to encourage elaboration of the integrated strategy for the coastal area in its narrower sense and the sustainable development of the whole town, thereby facilitating the best possible decision making for coastal area management.

4. Stakeholders and conflict resolution methods

4.1 Relevant stakeholders

- Ventspils local government, Ventspils City Council and its executive institutions:
  - Development Commission
  - Environmental Department
  - Architecture and City Building Department with territory planning office
  - Economy Department and Investment Department
  - Center of Tourism Information
- Neighbouring coastal local governments
- State environmental authorities at national and regional level:
  - State Environment Service and Ventspils Regional Environment Board
  - Marine and Inland Waters Administration
  - State EIA Bureau
- Non-governmental organisations – Environmental Protection Club, Latvian Environmental Education Fund and local society „Zaļā nagla" (Green Nail)
- Scientists and consultancies

In the COASTMAN project Ventspils City Council cooperates with neighboring coastal local governments because the environmental policy and management system cover not only the city territory but also all those areas around the city, which are of interest to and impacted by Ventspils and its inhabitants. Ventspils Free Port Authority and companies within its territory are also involved in the implementation of the project. Among the parties involved in the COASTMAN project are also state institutions – the State Environment Service's Ventspils Regional Environment Board and the Marine and Inland Waters Administration who control environmental protection and nature resource use in the land area, the continental shelf and the economic zone of the Baltic Sea in the Republic of Latvia.

- Ventspils Free Port Authority
- Harbor Master's Office and
- Private and public companies within the Free Port Authority territory

The Free Port of Ventspils Authority takes an active role in ensuring environmentally sound practice in the port through the following activities:

- Around the clock monitoring by oil spill collection team aimed at prevention and collection of incidental pollution in port waters. The unit is equipped with 6 oil skimmers and experienced personnel.
- The development of an oil spill contingency plan in cooperation with the port's terminals, where information procedures, risk assessment, response activities, and the application of available technology are outlined.

- The implementation of the Port's Waste Management Plan where information for port users is provided concerning different waste types (oily waste, sewage, solid domestic waste and other), treatment options, the capacity of reception facilities, reporting procedures, the responsibilities of organisations etc.
- The completion of the environmental infrastructure project by constructing an oil spill combating equipment storage facility at the northern breakwaters. The benefits of this project include the efficient deployment of different types of booms to locally confine pollution, and safeguarding against the spreading of oil into the open sea, thus minimising the impact on marine resources.

Navigation safety is also important for the guarantee of a quality environment in the port. These issues are under control of the Harbor Master's Office, which is one of Ventspils Free Port Authority's units. The Harbor Master's Office is working with state-of-the-art vessel traffic control systems, e.g. a modern radar system *STN ATLAS 9730* and radio-locator. 17 professional pilot crews ensure vessel traffic, and *Ostas Flote* LTD. manages 8 tugboats with an average 2500-3000 HP capacity.

Photo 2: Ventspils harbour

Photo: *A. Popeluha, 2004*

Taking into account that the River Venta is the only maritime access to the port of Ventspils it must be regularly dredged. The Free Port of Ventspils Authority guarantees a minimum depth for sea going vessels along the trajectory towards the Port of Ventspils. As there is a constant process of sedimentation, maintenance dredging is required year round. In the Free Port of Ventspils area of the River Venta the dredge spoil consists of fine sand mixed with a varying volume of mud depending on the season. In order to maintain the dynamic equilibrium of the river, the annual volume of dredged material averages 305.000 m^3. Over the last years the dredged material has been dumped in a special dumping area situated 4 nautical miles from the northern breakwater and 1.7 nautical miles from the shore line. According to Latvian national law regular checks and analyses of dredge spoil samples have been carried out accurately.

Throughout the year in the Free Port of Ventspils, environmental protection is highly valued and organised according to overall environmental protection rules and ethics.

Ventspils Port provides various kinds of environmentally important services including the removal and the utilisation of different types of waste that are polluted with oil and oil products, the removal and the purification of polluted waters from the washing of cargo tanks, and the removal of polluted ballast waters. The treatment of waste collected in the port is organised according to the rules and regulations of HELCOM, MARPOL 73/7 and Directive No. 2000/59 issued by the European Commission.

Ventspils port has realised the project „Environmental Protection against Oil and Oil Product Spillage in Ventspils Port". As part of the scope of this project a boom warehouse was built on the northern breakwater at berth No.36. Additionally, three new types of sea booms were bought: heavy sea booms (3 spools), light booms (6 spools) and absorbent booms (2 spools). The heavy sea boom spools can be operated with a diesel engine. The light boom and absorbent boom spools can be operated manually. In order to be able to use the booms in case of oil or oil product spillage a slip in the front of the boom warehouse was built. The slip helps to launch the booms into the water or to lift them out of the water. Additionally, to make the booms completely operational a mooring pile for boom strengthening was installed.

Today, as a result of the project „Environmental Protection against Oil and Oil Product Spillage in Ventspils Port" in the case of oil or oil product spillage it is possible to contain an oil spillage with absorbent booms as fast as in 15 minutes. The containment of the spillage with absorbent booms is particularly important due to the fact that after these preliminary works it is possible to close off the polluted water area with effective light booms or even with sea booms and skim the oil.

Photo 3: Storehouse for booms

Photo: V. Zilnieks, 2006

4.2 Legal methods: Environmental Protection, Planning, Building Control, Strategic Environmental Impact Assessment and Environmental Impact Assessment

On May 1, 2004 Latvia introduced the environmental impact assessment of planning projects also called *Strategic Environmental Impact Assessment* (SEIA) in compliance with the requirements of European Parliament and Council Directive 2001/42/EC of June 27, 2001 „On the Assessment of the Effects of Certain Plans and Programmes on the Environment". Its purpose is to identify and evaluate potential environmental impacts of certain plans and programmes and to involve the general public in discussing planning documents and decision-making, as well as to express proposals to prevent or decrease any potential negative impact. The strategic assessment is carried out during preparation of a planning document, before the document is submitted for approval.

This procedure was applied to Ventspils City Planning Document (2006-2018) in accordance with Decision No. 41 of February 25, 2005 of the national EIA office, although the development of the planning document was already in the development stage (its development was provided for in Decision No. 94 of February 23, 2003 of Ventspils City Council). After receiving the programme from the office, a consultancy was commissioned to prepare an environmental review. The environmental review should contain:

- goals of the planning document and its relation to other planning documents (Development Strategy)

- procedure of the strategic EIA, institutions taking part, public participation and results of this participation
- description of the current environmental situation and possible changes if the plan or programme will not be realised
- environmental situation in the areas that will be affected by the plan or programme
- environmental problems in protected areas
- relation to national and international environmental protection goals (National EPP, Baltic Agenda – 21, HELCOM recommendations etc.)
- evaluation of the main influences (direct and indirect, common effects, short- and long-term effects, positive and negative)
- solutions for minimisation of negative effects
- trans-boundary effects
- description of methods and principles used

In spring 2006 the consultancy firm finalised the Environmental Review. The main conclusions are:

The planning document determines opportunities, directions and limitations of the development of municipal areas, as well as their current and planned (permitted) use. Its main task is determining the planning of the town's development, taking into account the interests of inhabitants and businessmen ensuring both preservation of valuable natural resources and elimination or reduction of environmental pollution. The planning document of Ventspils City has been developed for the time period up to 2018. It reflects the long-term development trends of the municipality in this period, determining the current and the planned use of the area, as well as prospective research areas (construction of the northern port), for which opportunities of use will be researched in the future. In addition, the planning document defines the protective belts and also envisages development of infrastructure including a centralised water supply and sewerage network, heating supply, streets and bridges.

The planning document establishes the following changes in the current types of use of the territory and their development:

- to update the width of the Baltic Sea coast protective belt in the south-western and northern part of the city;
- to determine the protective belts of Lake Busnieku and the River Venta;
- to change the type of land use from agricultural and forest areas to industrial area in the north-eastern part of Ventspils in the Free Port territory;
- to improve transport infrastructure (reconstruction of the current bridge and of the port access roads, expansion of the airport, optimisation of free port territory etc.);

- to develop Ventspils free port activities by constructing a custom-designed coal terminal, fish uploading and refrigeration terminal, as well as other industrial objects;
- to develop a centralised water supply and household sewerage wastewater collection system;
- to develop a surface water, rainwater and ground water collection system;
- to improve the centralised heating supply system.

As part of Ventspils City's planning document a general description of the territory and environment was prepared featuring the following areas:

- air quality
- water consumption
- waste management
- ground and ground water pollution
- safety and risks
- green fields.

4.2.1. Possible environmental impact

It is envisaged in the planning document to change the width of the Baltic Sea protective belts in the south-western part of Ventspils, as well as the width of the protective belts at Lake Busnieku and the River Venta, and in the territory of Ventspils Free Port to change the type of use of a particular land section from forest and agricultural land to industrial territories in the north-eastern part of Ventspils. These areas were described in the environmental review including potential environmental impacts due to the implementation of these changes. It can be concluded that:

- Transforming valueless and overgrown agricultural land would improve the situation from the view of landscape, architecture and land use.
- Transformation of forest land in separate regions could aggravate air quality or at least keep air pollution at the current level. However, as a result the value of land would increase and possibilities to use and develop the area would improve.
- Changing the boundaries of the Baltic Sea protective belts would ensure structuring of the grey dunes and foredunes and coordination of the visitors flow.
- The planned construction of apartment houses in the area of Vasarnicu Street could adversely affect this area, as the number of people would increase, as well as the anthropogenic load on the forest and dune ecosystem of the seashore.

4.2.2 Activities to soften the impact

The softening of potential impacts and their evaluation will be provided through an environmental monitoring system including both the regular planning of the environmental policy and actions and evaluation of its efficiency and environmental licensing of economic activities. Currently the third Environmental Policy Plan and its corresponding action programme are in force.

To compensate for a possible reduction of the city's green area on the right bank of the River Venta, any future construction plans are to include large proportions of free areas for green spaces. Of particular significance is the establishment of new green areas (parks and squares) between the planned industrial and residential areas.

When planning the management of the Baltic Sea coastal forests, any excessive trimming should not to be allowed in order to avoid strong wind influence.

When defining the protective belt of the Baltic Sea coast and planning usage and management of this area both recreational and environment protection interests should be taken into consideration.

To preserve the dune ecosystem walkways need to be established to cross grey dunes and foredunes, as well as an access way for service transport. No mode of transport, e.g. pedestrians, horses, bicycles, should be allowed outside the designated footbridges. In the dunes located to the north of the gap opposite Seaside Park, recreation areas (fireplaces, tents, tables etc.), sport fields, sightseeing towers are forbidden, as well as changing the relief, forestation and planting.

To save the dunes, particularly the grey dunes; it is necessary to elaborate the management plan and methodology and to evaluate changes.

To ensure protection of the valuable habitats and protected plants in the Baltic Sea coast protective belt it is necessary to implement the solutions and recommendations as outlined in the section on regulation of the flow of visitors: closing of public roads, building walkways. When planning any development in this area, the use of plants foreign to the flora of Latvia and this specific habitat should be avoided. In no case should dry meadows be forested, since this would degrade a biologically diverse, natural and protected habitat.

4.2.3 Monitoring of implementation of Ventspils City Planning Document (2006-2018)

Elaboration of the planning document and its implementation are related to the Ventspils City Development Strategy for 2005-1015 and the implementation of Ventspils Environmental Policy Plan and its action programme. Within the Ventspils Environmental Policy Plan and its action programme periodic (every 3 year) evaluation of the implementation of actions of the programme and their efficiency are envisaged, closely linked to the monitoring of Ventspils spatial planning, as it would allow an evaluation of the affect the planned and implemented actions have had on the environment, including the green area in the town. To perform this evaluation data from monitoring, observations and statistics are used, as well as the results of research carried out by various experts, to develop a comprehensive picture of the changes of the environmental condition in the town and the efficiency of the implementation of the Environmental Policy Plan and the action programme.

On 22 May, 2006 Ventspils City Council decided to issue the final draft of the planning document for public consultation. Until 10 July the City Council received conclusions, proposals and recommendations from public institutions, as required by several legislative acts. Until the end of October four meetings were held with residents, who had submitted applications. Other residents or their representative groups were invited to meetings with the Architecture and City Building Department and the Development Commission. In addition, discussions took place in other municipal institutions.

4.2.4 Environmental impact assessment (EIA)

Environmental Impact Assessment (EIA) is a useful tool for the reduction of impacts on the environment. EIA is a multi-level procedure, which is required prior to the construction of significant facilities that may have a harmful impact on the environment. This procedure includes a set of measures, which envisages the review of the state of the environment at the given site, the review of impacts of the facility on the environment, the preparation of proposals for reduction or prevention of negative impacts, as well as the development of necessary requirements for the monitoring of the remaining impacts. In order to receive permission for commencement of envisaged activities, the Law on Environmental Impact Assessment requires that the applicant submit a final conclusion to the respective state institution or municipality. In addition, a report on such final conclusion by the authorised institution needs to be submitted, together with any other documentation as required. The municipality then carries out a comprehensive assessment of the final conclusion and the report of the national EIA Bureau taking into account the opinion of state authorities concerned and the general public before finally reaching a decision on acceptance or refusal of the envisaged activities. This decision is then displayed in the local municipality building and in other public places, as well as being published in the local newspaper.

In Ventspils, EIAs have been carried out for a number of reconstructions in the largest oil and product terminals in the port territory, as well as for the construction of a number of new production facilities and a sewage water treatment plant.

In cases where according to the law an EIA is not required, the Ventspils Regional Environmental Board has been carrying out initial EIAs, and only thereafter technical requirements for design and construction works are being issued.

4.2.5 Local municipal legislative methods by laws (including planning and building control) and licenses

On a local level, relevant legislation includes municipal binding regulations. In case of Ventspils, the Environment Protection Licensing Regulations, Seashore Management Regulations and Building Regulations, which are currently under development, should be mentioned.

A successful instrument for implementing environmental policy in relation to companies is the environmental licensing system that has been in force in Ventspils since the mid 1990s. Ventspils wanted to provide for the realisation of the functions and rights given to the local authority by the state. In the mid 1990s the environmental legislation in Latvia was comparatively disjointed, but in many countries, e.g. the Netherlands, it is more coherent. That is why the principles of environmental licensing as enforced in the Netherlands have been used as the basis for Ventspils' environmental protection licensing regulations. The licensing system means that every enterprise or activity, which may have an impact on the environment (either through emission of pollution or disposal of waste) and/or may create any risk for the environment, must receive an environmental license – a special permit from the municipality. The process of preparation of the license includes the following main stages: application, informing the general public and public hearing, preparation of a draft license, preparation of the license and issuing the license.

Thus, the licensing system provides stricter environmental standards than the national requirements, a possibility to forbid some activities of companies if they are considered to be dangerous or hazardous to environment and population of the town, taking into account the public opinion. Public hearings are the main measure to develop democratic principles in our society. Therefore, Ventspils has already organised its work according to the EU Directive on Integrated Pollution Prevention and Control (IPPC).

The local authority compiles the draft license taking into consideration Latvian environmental legislation norms, the Ventspils Environmental Policy Plan, the Ventspils Territory Development Plan, the conclusions of the national Fire and Emergency Service, the national Public Health Agency, the national Labour inspectorate and the national environmental inspection, as well as objections expressed during the public hearing.

The license contains environmental protection and safety regulations and requirements, as well as the requirement to carry out measurements in order to trace environmental changes and the level of pollution, and inform the local authority about the results of the measurements.

In certain issues the City of Ventspils, by implementing its environmental policy, has been ahead of the processes taking place on the national level. For example, the requirement to install vapour collection systems in newly built filling stations has been in force in Ventspils since 1994. Similarly, the requirement to carry out a soil and groundwater contamination study before starting construction works was incorporated in all the environmental licenses issued to enterprises.

The control of compliance with the license requirements is performed by the local authority, and when deemed necessary the State authorities that have participated in developing the license can be invited.

4.3 Environmental information collection and research methods

4.3.1 Research on curret geological process of the seashore

During the COASTMAN project Ventspils municipality is continuing its cooperation with independent consulting companies and scientists, mainly to investigate the value of natural resources and to work out the forecasts of changes in nature that have to be taken into account in the planning and building in new areas.

This includes the current investigation into the *geological processes of the seashore*. Problems with changes in the coastal zone are common to almost all of Latvia's seacoast. Erosion processes are mainly natural, but there is also the influence from the port. The port's long breakwaters and up to 17m deep shipping channels practically cut off alongshore sediment drift and lead to significant changes in the coastal belt on both sides of the port. To the north of the port downdrift erosion is continuing due to the alongshore sediment deficit. This zone is 10-14 km long and includes a belt of land loss with a width of 70-200m to the south of the harbor structures. Over the past century since these harbour structures were built updrift acceleration has been observed in a 5-10 km stretch, attaining a width of 300-600 m with a maximum of 900 m.

Taking into account such processes in the coastal zone, it is vital to have a qualitative forecast of changes in order to make plans for future development, such as building and planning of recreation activities on the seashore.

For planning of the region it was necessary to take into account natural processes such as coastal erosion in the northern part of Ventspils. The municipality was eager to obtain proposals for recommendations and forecasts and therefore requested scientific assistance.

In order to understand the geological processes in the northern part of the Ventspils region, one particular district can be used as a model. According to the relative elevation, morphology and geological structure three different coastal sections, each approx. 360-380 m long, can be named.

Fig. 1: Map of the seacoast in northern part of Ventspils with the height of bluff in meters

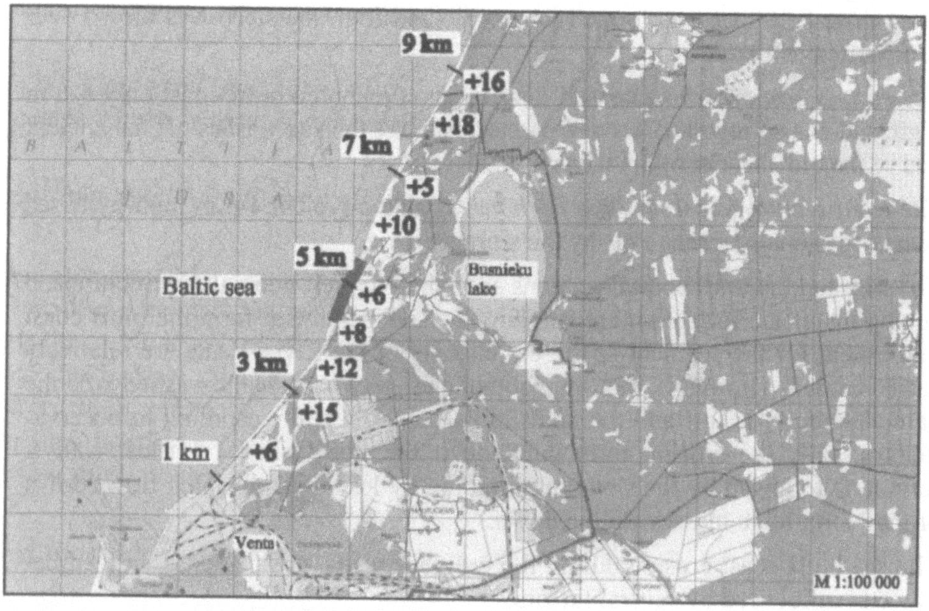

Source: G. Eberhards, 2005

Fig. 2: Detail analyzed district of the bluff

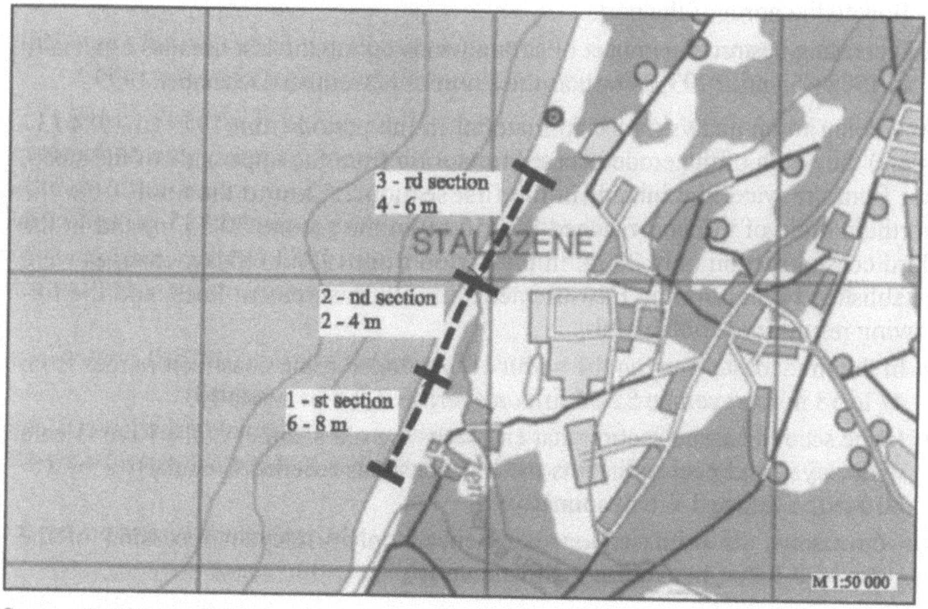

Source: G. Eberhards, 2005

The first section (the most northern one) is a strongly marked 6-8 m high bluff coast regularly eroded by storms, with an almost fully eroded 2-4 m high narrow coast dune belt, that was formed when the sand from the steep was blown over its upper edge inland.

The second section, a low coastal district, is an indented bluff coast only 4-6 m high, and is also regularly eroded by storms. Here the sea washes off the ancient bed of the old channel of the River Venta.

The third section (closest to the harbour) is also a typical 4-6 m high bluff sea coast occasionally washed off by the storms.

The first and the third coastal sections according to their geological structure and conditions of the formation are similar. The sediments that form the bluff coast belt except for the segment of the old channel of the River Venta are relatively easily eroded. During storms the foot of the steep is usually washed off, the natural stability is lost and the landslip processes begin. According to observations, usually the erosion from the foot of the steep when a vertical wall is formed is followed by the slope retrograding by losing sand, which flows down or may form large masses sliding down.

The sea bluff coast erosion during the previous years, as detailed described below was influenced by several factors:

- The proximity of port waterworks – breakwaters and continued (1953-1993) dumping of dredged material from the waterway channel at the distant sea dump, which created up to 30 million m^3 drift deficit in the alongshore drift flow to the north of the port.
- Increasing destructive impact of strong winds during the last decades especially in 1982-85 until 1993, as well as the storm of November-December 1999.

According to an analysis of map material in the period from 1947 to 1984 (37 years) the width of the eroded coast (measuring from the upper side of the steep, not from the waterline inland) in the first section 2.5 km to the north from the northern mole of the port reached 75-150 m, in the second 30-35 m, and in the third coastal section only 20 m. In the period from 1982 to today, changes were established by performing measurements in 20 measurement lines, and the following results were obtained:

- In the first coastal section the width of the eroded basic coast belt ranges from 47 to 58 m (on average 52-55 m), on average 3-3.5 m annually.
- In the second coastal section that embraces the old channel of the River Venta with clayey and peat sediments the low coast has receded irregularly – by 18-30 m (on average 1-1.6 m annually)
- Conversely, the third section experienced only irrelevant recession of the steep by 3-9 m, (on average 0.5 m annually).

It is expected that in the following decades the dominating wind direction will not change, that the destructive activity of the wind will increase, and that the drift deficit in the shallow waters of the sea to the north of Ventspils port will remain.

Taking as the base the long-term average of erosion and recession speed as main tendencies, we can forecast that in Ventspils' northern coastal part, where for half a century substantial erosion of the terrestrial belt was observed, in the following 50 years (*until 2050*) the width of the actually endangered basic coast belt (actual risk zones) will be as follows: 1st section – 90 to 100 m; 2nd section – 40 to 60 m; 3rd section – 30 to 40 m. However, if in *2020-2030* as forecasted on a global scale an increase in temperature will take place, including an increase in the frequency and strength of storms, especially of those regional south west- westerly winds, then we can assume that the coastal geological activities will increase and the width of the belt that is being eroded as well. In the maximum possible erosion risk zone it may reach 140-150 m. This leads to the following question:

What should be done when as a result of erosion of the main coast a large number of measured land plots intended for individual land owners or tenants will be lost?

Experience from around the world shows that depending on the significance of the endangered coastal segment (e.g. apartment houses and infrastructure of the town, port buildings, unique coastal landscapes and beaches significant for recreation, or underpopulated or little used, swampy, covered with bushes, areas with degraded environment etc.) on a local or national scale, nowadays there are three approaches (three options) bearing in mind that protection against erosion is very expensive:

- *1st option is not to do anything* if an area is of little significance, scarcely populated, has no significant natural resources or excellent protected nature objects or cultural and historical monuments. People need to adapt to the changes taking place in the coastal belts.
- 2nd option is to retreat further inland.
- 3rd option is to actively protect it and build nourished beach (replenishing a beach artificially by delivery of materials dredged or excavated elsewhere) or build coastal protection constructions (booms, breakwaters, walls, rip-raps etc.). At the same time it must be mentioned that over the last decade leading experts around the world have advised against the construction of heavy, massive waterworks (booms, breakwaters, rip-rap, etc.), which are very expensive to design, construct and maintain and not always provide positive results. The other significant reason against construction is that the natural coastal belt, its environment and landscape would be destroyed. Often, so called soft constructions take their place, environmentally friendly technologies and methods that help to preserve the natural look of the coastal belt: regular renewal of eroded beaches, drifting substantial masses of sand, draining of beaches etc.

Photo 4: The lonely tree on the bluff

Photo: I. Plavina, 2005

The main problem is where to source sand with the necessary granulometric content (roughness), as it must be at least as rough as in the beach, which is to be worked on, but preferably rougher. Beach nourishment is not a once off but periodic measure taking place every 3-10 years, depending on how fast strong storms wash the sand into the sea. To eliminate or reduce catastrophic coastal erosion in a 1.1 km long and approx. 50-60 m wide segment the beach needs to be drifted (i.e. made 2-3 times wider than the existing one), and the amount of sand needed is 100.000-150.000 m^3. Building booms or other waterworks systems in the sea and on land that degrade the natural landscape would decrease or even eliminate erosion of the mainland, but the erosion would become even more intense to the north of the protected coastal segment. Building waterworks for the protection of coast would be more expensive than renewal of the beach and widening and elevating it.

Obviously, in the northern part of Ventspils' seashore the 3rd option is not recommended, neither from the point of view of preserving natural coastal environment, nor from an economic point of view, but the first one is advocated– not to do anything.

4.3.2 Research on urban forest quality and impacts

Another investigation of natural resources and their features carried out in cooperation with scientists is the *investigation of forest health and the impact of the city.*

In the administrative region of Ventspils City there are considerable areas with preserved natural ecosystems – mainly forests – 1648 ha, or 30% of the total city area. Forests are mainly located in the northern part of the city along the Baltic Sea coast and around Lake Busnieku. The forested land reaches out into the industrial area of the city. It consists mainly of pine (87%), with the majority of the trees being middle aged. In the southern part of the city black alder woods occur. Ventspils City's forests are ranked among the outstanding national forest areas, with the highest cadastral value (154) in the country. Ventspils City's forests are multifunctional. They perform a sanitary function – intercepting polluting substances and noise, and a recreation function – serving as the environment for active recreation of the population. The forests play a great role in the regulation of the microclimate of the city, in anti-erosion measures, in preservation of the environment, as well as in biological diversity. The forests of the city are an integral part of the urban scenery and serve an important social function. The forest in the dunes and the direct proximity of the sea create a beautiful natural scenery for people to enjoy the peculiar forest microclimate, silence and tranquility.

Urban forests are subject to intensive anthropogenic loadings that adversely affect their quality. In the forests with a high recreation load it is very difficult and near to impossible to protect the disturbance of sensitive species. Due to increased trampling the soil is being compressed and the wood growth decreases. In the regions adjacent to construction areas the soil is subject to eutrophication, and synantrophic species appear in the ground vegetation and undergrowth due to human activities. In certain localities deciduous trees appear among the typical pines. As the load created by the holiday-makers in the natural habitats in the area of the city often exceeds the optimum, the habitats have significantly changed and their recreational value has decreased. Forest fractioning caused by construction works also is a negative aspect. The forests are being polluted with household waste, causing change not only to the structure of species in the forest stand, undergrowth and ground vegetation but decreasing its aesthetic value and naturalness, and reducing the ability of the forest stand to perform air and soil protection functions as well.

Plants can react very sensitively and dramatically to changes in the environment. The framework and structure of the forest ecosystem are closely linked to the environmental factors. Forest ecosystems belong to the terrestrial ecosystems that significantly change due to changes in environmental conditions and pollution; therefore both the living organisms and the whole forest ecosystems are sound indicators of the state of the environment. By using the indicator ability of the forest stand and organisms in certain locations of the forest it is possible to evaluate the overall environmental condition of the area.

Photo 5: Oil terminal surrounded by forest

Photo: *„RHINO dizains", 2003*

In order to investigate the impact of the city environment on the forest ecosystems the monitoring of the forest ecosystem condition began in 1993 in the forests of Ventspils under the supervision of Prof. M. Laiviņš (University of Latvia). In the natural forest stands located in the area of the city eight permanent forest monitoring sampling plots were established. They are located at varying distances from industrial objects and residential areas, wherewith anthropogenic load of the different forest stands differs. For data comparison five sampling plots were established outside the city in the forests of Ventspils' region. Local forest monitoring in Ventspils was carried out:

- To obtain data on the framework and structure of the forest stand in the different locations of the city;
- To establish by comparing and analysing the obtained data what changes are taking place in the forest stands due to the human activities, including pollution, to describe the types of changes, to classify them, to attempt revealing the causes of the changes and to forecast the state of the forest stands in the future;
- To evaluate the environmental condition of the city on the grounds of the structure of the plants in the forest stands and peculiarities of wood formation.

At the sampling plots data is collected according to the main parameters of the forest stands – woody plants, herbaceous plants, moss, lichen and soil.

The best indicator for the health condition of woody plants is defoliation of the crown directly reflecting the quality of the growing conditions. The prevailing tree species at the monitoring sampling plots is pine. Changes to the tree crown are caused by both air pollution and strong wind damage, as well as influence of vermin and diseases.

The monitoring of crown health shows that defoliation (loss of needles in the crown) of pines in the forests of Ventspils City is greater than in the forests outside the city. In the forests of the suburbs the fluctuations of defoliation over the years are negligible and generally invariable; however, in the forests of the city a slow but constant defoliation increase is being observed, consequently the health condition of the trees is worsening.

A clear indicator of the air pollution is the number and projective cover of lichen and alga species. The investigation of epiphyte species (lichens and algae growing on the trunk) at the forest monitoring sampling plots shows that the number of species and the projective cover is higher in the sampling plots further from the city and to a lesser extent influenced by human activity. In the forests of the city those sampling spots that are located in the areas under dominant wind influence posses less epiphyte species. All this presents evidence for a greater amount of polluting substances in the air of the city than in the suburbs.

From the investigation of vascular plants and moss in the forest monitoring sampling plots during the 12 year observation period it has been possible to establish a trend for the number of species of the vascular plants and moss to increase. This also clearly shows the changes in the growing conditions due to anthropogenic influences. Usually, the soil of a pine forest is nitrogen poor. When the soil is being fertilised species appear, which are typical for industrially affected areas.

Overall, the health of the forests in Ventspils does not differ from the average Latvian level. Changes in the forests of Ventspils City and its suburbs are not rapid, but rather slow and gradual.

4.4 Voluntary information and public awareness methods

4.4.1 International certificates for the Quality Management System ISO 9002 and the Environmental Management System ISO 14001

The largest port companies have international certificates for the Quality Management System ISO 9002 and the Environmental Management System ISO 14001. All of the port companies have obtained permits regarding pollution activities issued by the State Environmental Service's Ventspils Regional Environmental Board and environmental licenses issued by Ventspils City Council.

4.4.2 Blue Flag campaign and cooperation with NGOs

Coastal management is promoted by various voluntary movements as well. As early as 1997 Ventspils City local government decided to get involved in the Blue Flag movement. Since 1999, each year the *Blue Flag* awarded by the by the Foundation for Environmental Education (FEE) has been flying over Ventspils beach, showing not only development of bathing places and cleanliness of bathing waters, but also a systematic and purposeful approach to solving environmental issues in local government.

Initially, the Blue Flag campaign was a measure to motivate local government institutions to ensure clean and safe beaches and marinas for local residents and tourists. Gradually, the campaign has transformed into a wider one in the context of coastal management. From developing separate beaches the campaign has expanded to defining municipal policies for coastal environment development promoting sustainable use of coastal resources, planning of coastal zone management and improving water quality. Substantial changes in the way of thinking have taken place recognising the meaning of integrated coastal management. In Latvia the Blue Flag programme is being implemented since 1998 when the Environment Protection Club, a Latvian NGO, joined the Blue Flag movement as a national coordinator.

Photo 6: Ventspils beach

Photo: *G. Seilis, 2006*

The Blue Flag campaign requires meeting of 28 criteria – the main ones being the high quality of bathing waters and organising environmental education activities. The mentioned education activities in Ventspils have been organised together with the city's Educational Board as competitions on different environmental issues for schoolchildren. Testing of bathing waters is carried out according to state regulations. Samples for bathing waters quality must be taken twice a month at three different points on the beach. These regulations also determine quality indices and desirable and obligatory requirements. Unfortunately, these regulations do not conform to the requirements of the Blue Flag campaign, so it is necessary to carry out additional tests in order to certify the good quality of bathing waters. Because of the main streams that come from the south-west and the location of the sewage treatment plant to the north of the town there are no threats to the bathing waters. The only potential threat is oil, which could come from oil terminals in other countries to the south of Latvia and from ships in case of accidents at sea. Ventspils port is separated from the sea by the breakwaters thus making any potential pollution from the port to the sea near to impossible. In the case of any leakage it is possible to close the sea gateway with the help of booms, and polluted waters can be gathered inside the port.

5. Possible solution and lessons learned

An evaluation of the methods and tools used in organising Baltic Sea coastal zone management and their efficiency will enable research and protection of valuable natural resources in the coastal zone. In turn, analysis of the role of local government in the environmental evaluation procedure will allow full implementation of functions of the municipality prescribed by the legislation. Thereby local government has the potential to implement the goal of the environmental policy – efficient environmental management on behalf of Ventspils' inhabitants, ensuring economic growth and following the principles of sustainable development.

A document such as the Environmental Policy Plan is used by the city as a flexible environmental management tool, which allows it to promptly make informed decisions on effective solutions for various environmental problems and to ensure continuous monitoring of the implementation of the environmental action programme. The plan facilitates negotiations with big companies, encouraging them to be more active in the development or improvement of their environmental management systems, further reduction of the risk of accidents and reduction of pollution. The plan also provides for the dissemination of more comprehensive information to the public on the environmental status of the city and encourages greater public involvement in decision making on environmental issues.

Participation in the Blue Flag movement is a good method for promotion of beach management and environmental education.

The introduction of the strategic environmental impact assessment procedure will ensure that the developers of different planning documents (development plans, strategies of economics and its different branches, as well as regional development plans) will be responsible for the evaluation of how the planned development might influence the environment and that the general public can get acquainted with the draft document and the included environmental review and express their opinion.

Photo 7: Storage tanks of oil products and trees in blossom

Photo: A. Popeluha, 2004

6. Conclusions from the case

Coastal zone management must be organised by implementing a large number of tools and methods on a national and local level, especially spatial planning, as well as by involving all interest groups- municipal and local institutions, developers, enterprises, NGOs, independent consulting companies and scientists.

The most useful instrument for planning not only in the coastal zone but in all the areas of this city located on the coast of the Baltic Sea is strategic environmental assessment, which is carried out during the preparation of the planning document, before this document is being submitted for approval. The strategic impact assessment is to be carried out regarding the planned development, implementation of which can substantially influence the health of the population and/or the environment, including developments in the areas of agriculture, forestry, fishery, energy, industry, transport, waste management, water resource management, telecommunication, tourism, mining, as well as plans related to regional development, land lease, spatial planning and the European Union funds.

Due to the fact that the port plays a dominant role in the development of Ventspils, efforts by environmentalists and planners are devoted to the following aim: the development of economic activities in a free port area while at the same time preserving valuable natural resources is possible only by following the principles of sustainable development – development that meets the needs of the present generation without compromising the needs of the future (Brundlandt report).

7. References

Annual reports of Marine Inspection. Available at: www.jvp.gov.lv/gada_parskati.htm

Cabinet of Ministers Regulation No 157 of March, 2004. Procedures for Strategic Environmental Assessment. Available at: www.vidm.gov.lv/varam/Elikumd.htm#ivn

CCB report „Checking coastal conservation. International requirements to safeguard nature on the Baltic Sea coast" (2004), Uppsala, Sweden. Available at: www.ccb.se/downloads/ccb_check05.pdf

Eberhards, G. (2003). Latvijas jūras krasti, Latvijas universitāte, Rīga.

– (2004). Jūra uzbrūk. Ko darīt? Latvijas Universitāte, Rīga McĀbols Available at: http://piekraste.daba.lv/LV/izdale/Jura_uzbruk_Ko_darit.pdf.

Gronholm, B., Joas, M. & Nordstrom, M. (2001). Success Factors for Local Agenda 21 in the Baltic Sea Region, Abo Akademi University. Available at: http://lu.se/IIIEE/EMAS_Eastwards.

Latvijas veģetācija. 5.sēj. Latvijas Universitāte ģeogrāfijas un zemes zinātņu fakultāte Bioģeogrāfijas laboratorija, Rīga 2001. Laiviņa M redakcijā.

Law on Environmental impact assessment. Accepted 13.11.98. Available at: www.vidm.gov.lv/ivnvb/ivnvb/likumd/Livn_lik.htm.

Manual of Ventspils Regional Environmental Board. Available at: www.rvp.gov.lv/index.php?dir=19/.

Meima, R. & Starkey, R. (2000). Implications of the Spread of Voluntary, Standardized Environmental Management Systems in Central and Eastern Europe. Final project report. Lund University, Sweden.

National Environmental Policy Plan accepted by Cabinet of Ministers on 03.02.2004. Available at: www.vidm.gov.lv/varam/DOC/vapp/Lnvpp04.htm.

Sauer, P., Andrews, R., Cudlinova, E., Dvorak, A., Fiala, P., Holcova, P., Kovar, J., Lisa, A., Livingston, M. & Tosovska, E. (2001). Voluntary agreements in environmental policy. University of Economics, Prague.

Strategy for Sustainable Development of Latvia approved by the Cabinet of Ministers of the Republic of Latvia 13.09.2002. Available at: www.vidm.gov.lv/varam/DOC/Eilgtsp_att.htm.

Ventspils Environmental Policy Plan. Popular version, Ventspils 2001-2005

Wackernagel, M. & Rees, W. (1996). Our Ecological Footprint. Reducing Human Impact on the Earth, Norden AB.

World Health Organization (2000). A Source Book on Implementing Local Environment and Health Projects. Chadwick House Group Ltd, London.

Chapter 4

Pilot Course on 'Conflict Resolution applied to Coastal Zone Conflicts in the Baltic Sea Region'

–

The Educational Part of the Coastman Project

Nils Brandt[1], Jan Fidler[1], Ronald Wennersten[1], Åsa Larsson[1]

Photo: *Arija Sproge*

1. Introduction

The Coastman project (Coastal Zone Management in the Baltic Sea Region) studied real and/or possible conflicts in a number of harbour areas in countries around the Baltic Sea (Estonia, Finland, Germany, Latvia, Lithuania, Russia, Sweden), with the aim of gathering knowledge and developing methods and solutions for conflict mitigation and sustainable planning processes in coastal ar-

1 Royal Institute of Technology (KTH) – Department of Industrial Ecology – Stockholm – Sweden.

eas. These studies took the form of case studies carried out in the different countries with the focus on conflict resolution, sustainable development (SD) and experiences from planning processes and stakeholder analyses.

As part of the Coastman project, a pilot course on 'Conflict resolution applied to coastal zone conflicts in the Baltic Sea region' was developed and implemented during the autumn of 2006. The aim of this pilot course was to provide deeper knowledge and understanding of conflict resolution in coastal zone conflicts in the Baltic Sea Region (BSR) from a SD perspective.

The basic idea when constructing the course was to collect and share experiences gathered in the case studies and to provide participants with the potential to exchange knowledge and experiences and to improve their practical skills and develop their understanding of conflict resolution methods. This was mainly achieved through studies and discussions of different methods, tools and experiences of conflict resolution around BSR.

The focus group that participated in the course consisted of project partners and variety of invited stakeholders from different countries around the BSR. There were about 35 course participants in total, with diverse backgrounds and experiences such as scientists, politicians, municipal officials, harbour officials, government employees and students. Some of these were directly involved in the project case studies carried out in different countries, while others were purposely invited to take part in order to create a desirable mix of experience and knowledge among course participants.

Scientists came from: Tallinn University of Technology, TuTech Innovation GmbH in Hamburg, University of Latvia, Klaipėda University, The Royal Institute of Technology, Commission for Nature Protection of the Russian Geographical Society, Russian State Hydrometeorological University.

Officials came from: Municipal Government of Haapsalu (Estonia), Riga City Local Agenda 21 Office (Latvia), SIA (Regionalie projekti, Latvia), Ventspils Port Authority (Latvia), Port Infrastructure Office (Lithuania), Primorsk Urban Municipality (Russia).

Government employees came from: Nacka Municipality (Sweden), Free and Hanseatic City of Hamburg Department, Environmental Emergency Response Unit (Germany), Ventspils City Council Environmental Department (Latvia), Territory Development and Planning Group Property and Territory Development Department (Lithuania), Stockholm City Planning Committee.

2. Background information

2.1 Pedagogical ideas behind the course

When constructing the course, the following pedagogical concepts and ideas were included:

- The pilot course was built around a virtual case study and a role play, both based on the same decision-making situation – where to build a new oil harbour. Using this approach, the course participants were given the potential to experience conflict resolution without having self-interest in the conflict.
- Role play was chosen as a suitable learning method because it provides the opportunity to reflect on the complexity of social or environmental issues and has the potential to facilitate learning processes in these fields (Henriksen 2004, p. 127).
- The pilot course was not constructed as a university course for higher education students; it was a course based on the knowledge and values of different types of stakeholders. This meant that a very important part of the knowledge and experiences came from the participants themselves, supplemented with lectures and discussions of different methods, tools and experiences of conflict resolution around the Baltic Sea Region.
- Due to the international character of the course, a website was developed to serve as an information and communication platform – providing study materials, course results, links and a discussion forum (www.webforum.com/ pilotcourse).

2.2 Course structure

The course was divided into two seminars held over a three-month period. The seminars included individual assessment, lectures, work in groups and discussions. Between the seminars the website served as a contact link. There were thus three essential components of the course: the virtual case study, the role play and the course website.

Seminar 1 began with a block of lectures, after which participants were divided into smaller groups to discuss what SD means in general and to attempt to reach consensus about the criteria that should be used to evaluate SD in a harbour planning process. These criteria were used to compare four harbour locations suggested in the virtual case together, with analysis of possible conflicts. The last task was to choose the best possible harbour location and suggest methods and procedures for the decision-making process. Results from the group work were documented and presented during the final joint discussion.

Seminar 2 centred on role play. The role play simulated a decision-making process, and the course participants played the roles of different stakeholders, practising negotiation and workshop techniques. The role play storyline was based on the virtual case. Two groups performed the same role play, discussing several topics and trying to reach agreement about each of them. At the end of the seminar, both groups reconvened and presented their results.

2.3 The virtual case

The virtual case study developed for the pilot course simulated the possible geographical, environmental, historical and political situation between two countries. To illustrate this situation, a storyline and a map were used. The storyline described the current situation between the two countries and the dynamics of their relationship, while the map visualised the storyline. Both the storyline and the map were presented in a way that left spaces for course participants to fill with their own ideas.

Fig. 1: The map used in the virtual study case with four suggested harbour locations (A, B, C, D)

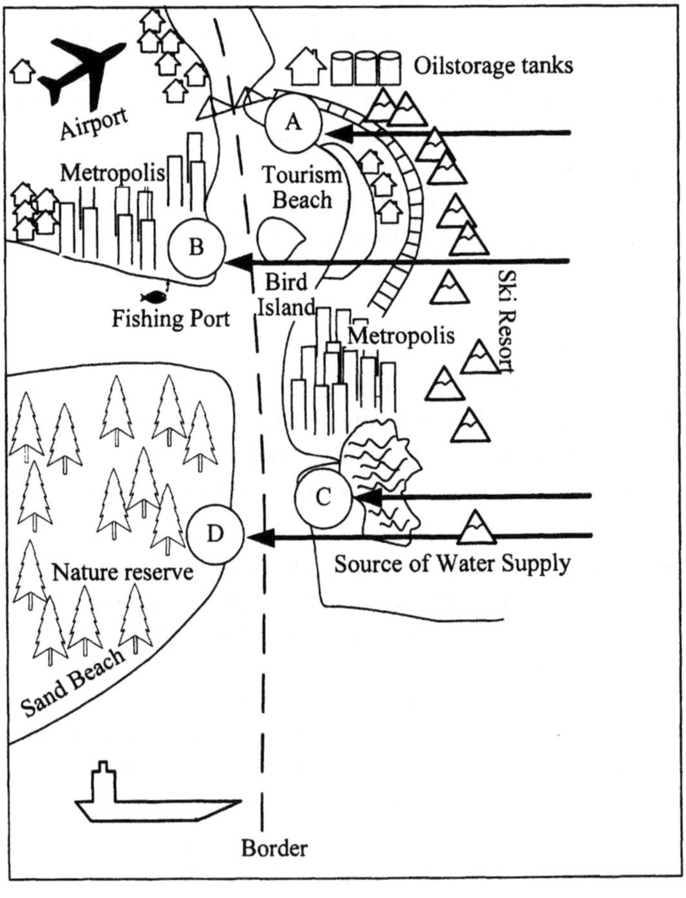

The storyline described two neighbouring countries that were both experiencing rapid economic growth and that had decided to collaborate in building a new port for oil imports to their two largest cities. The countries were situated by the sea and separated by a bay. A bridge with good transportation capacity connected the two countries. The coastline and the water were used for different purposes, for example tourism, fishing, water supply and natural habitats. Four different harbour locations were presented as possible locations (Figure 1).

This virtual case was used during both seminars but with different approaches. During Seminar 1 the main task was to choose one location as a possible future harbour from four suggested locations, after analysing the four locations according to SD criteria and possible conflicts. None of the suggested locations was an optimal choice. During Seminar 2 the same map was used but the storyline was rewritten in order to suit the purposes of the role play.

2.4 Website

As mentioned above, a website with different functions was developed for the course. Without these functions it would not be possible to carry out a distance learning course such as this.

The website was divided into two parts; a restricted site with detailed information for the course participants and a public site with a general description of the course and contact information for people interested in receiving more information about the course.

The following information was made available on the restricted site:

- Logistical information; course programme, timetable for seminars
- Virtual case description; map and storyline
- Material for the role play
- Library with study materials
- Material from the lectures
- Results from group work during the seminars.

The course website also contained useful functions such as:

- Information & exchange forum
- Sharing of documents and other materials
- Discussion forum
- Personal profile.

The course participants were introduced to the website by filling in their own profile and by uploading a short reflection about how they wished to relate their professional work to sustainable development (SD).

During the entire course, all information about course content, timetables and logistical information was distributed through the website. On the website, the course participants could find information about each other and keep in contact.

Before the virtual case seminar in Riga and the role play seminar in Stockholm, questions for the group work and role play instructions were published on the website. After the seminars, the course participants presented their group work and the results of the role play in the website library.

3. Seminar outcomes and lessons learned

3.1 Seminar 1: Riga. The virtual case – choosing the best possible harbour location

The seminar in Riga lasted one and a half days and began with lectures, after which the participants worked with the virtual case in groups. There were five working groups consisting of 6/7 individuals. The seminar finished with a joint plenary session where group work results were presented and discussed.

The main focus during the seminar was on sustainable development (SD) and conflict resolution. The work was divided into the tasks listed below and participants were asked to keep a record of their work in order to make presentations during the plenary session and to produce a written report.

- Write a personal reflection about how to relate your professional work to the concept of SD
- Discuss SD, develop and agree upon criteria for SD
- Create a table that compares four given harbour locations according to the criteria
- Analyse the four harbour locations according to possible conflicts
- Choose one harbour location as a result of your analyses; motivate your choice
- Suggest how the decision-making process associated with the chosen harbour location should be carried out and list the methods that could be used.

The first day started with a welcome speech by the Latvian Minister of the Environment, followed by opening information about the course structure. The subsequent block of lectures covered different subjects relating to the planning processes, stakeholder participation and conflict resolution and analyses. The first lecture was an introduction to a method of framework analysis, with the applications and experiences from studies in Guanabara Bay, Brazil. The next lecture informed participants about planning and decision-making processes in Swedish municipalities, with an example from Nacka municipality. The last two lectures of the first day were about ideas and concepts of SD and methods for conflict resolution and decision-making processes. A lecture given during the second day described the opinions of a Latvian NGO concerning harbour expansion in Riga. All lectures given during the seminar were later made available on the course website.

In order to prepare the participants for the course, they were asked to submit an assessment before the seminar. Participants were asked to write a short reflection about how they would like to relate their professional work to the concept of SD and upload it onto the course website. These reflections were also briefly presented by participants when the group work started.

The work in groups started with a discussion about concepts of SD; participants presented their personal opinions and the first task was to reach consensus about the criteria that should be used when evaluating SD in a harbour planning process. The next step was to analyse the suggested harbour locations in the virtual case according to the SD criteria. The participants had to choose one location that best fulfilled the criteria for SD. The last task was to discuss and devise suggestions regarding how to carry out a decision-making process in connection with the chosen harbour location and the methods that should be used for this.

Photo 1: Working in groups during Seminar 1 in Riga

Photo: Arija Sproge

The participants were asked to record all results and write a short summary of the discussion. The closing part of the seminar comprised a plenary session during which the groups presented their results and discussed them together. Finally there was an open vote where all groups decided the best location for the proposed harbour. After the seminar, groups had two weeks to complete their reports and upload them onto the website.

3.1.1 The results of Seminar 1

All working groups succeeded in completing the tasks assigned and chose a harbour location according to a suggested methodology. However the results differed between groups. This was a good result because the virtual case was created so as not to have one optimal solution.

During the work with the virtual case, participants found that the virtual case storyline was written in quite a general way. This approach created confusion in the beginning but also encouraged participants to use their imagination and initiated discussions in groups. Participants were asked to fill out the storyline with their own ideas and suggestions.

Fig. 2: The chosen harbour location according to the five groups. Two groups chose location A, two groups location C and one group location B. None of the groups chose location D

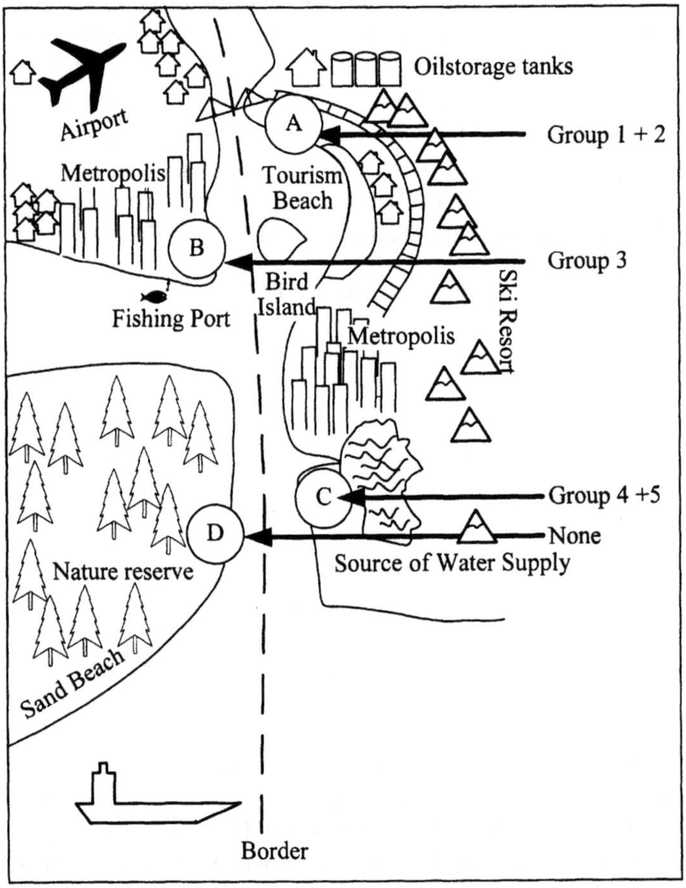

There were no disputes as regards SD consisting of three aspects; economic, social and environmental. The discussions arose when different groups had to decide what criteria should be used to evaluate harbour locations and whether the criteria should be positioned under the different aspects (economic, social and environmental) or handled all together. The argument for not grouping criteria under different aspects was that some of the criteria could be added to all groups.

After analysing the harbour locations according to the SD criteria and possible conflicts in all four locations, a consensus had to be reached about where to locate the new harbour. The results from the groups were not those anticipated when the virtual case was being constructed. When comparing harbour locations according to the SD criteria, groups used different approaches. Some groups made a more quantitative assessment, evaluating locations according to SD criteria using a grading system, e.g. 1-5 or labelling locations with plus and minus points and then adding these together. Two groups completed their analysis with additional criteria: one group suggested worst case scenarios and another group legal circumstances as additional criteria for the final harbour choice.

After choosing a harbour location, the groups gave examples of decision-making processes that could be used, e.g. SWOT analyses, logical framework analyses, hearings, interviews, downstream conflict resolution processes. At the end of the seminar all groups gathered together and presented and discussed their results in the final plenary session. In the open vote about harbour location, locations A and C got the most votes, followed by location B as the best for the future harbour. None of the participants voted for location D.

3.2 Seminar 2: Stockholm. The role play – decision/negotiation process simulation

The seminar in Stockholm lasted two days. Participants performed the same role play in two separate groups, each consisting of over 15 participants. The role play simulated a decision-making and negotiation process which was divided into two different parts – a pre-negotiation part and a formal negotiation. The participants acted as stakeholders negotiating about the status and responsibilities of a future harbour commission.

The role play storyline was based on the virtual case, rewritten for this purpose. The storyline now concerned a negotiation process in which two countries agreed that a Permanent Harbour Commission (PHC) should plan a new oil harbour. Before the PHC could start its work, a Temporary Commission (TC) had to meet to discuss and decide on several issues that were essential for the PHC work.

The central issue for the whole negotiation process was to reach agreement about the following topics, which were chosen to reflect the aims of the course:

- Procedures of PHC (members and chairman)
- Guiding principles of the PHC
- Defining mandate of the PHC
- Securing financial resources for harbour construction, location and impact
- Defining procedures for establishing a broad support in the process of finding and deciding the new harbour location
- Assessing environment consequences of the new harbour.

The TC consisted of the following stakeholders:

- Ministries; trade and environment
- Small Business; tourism organisations, smaller fishing industries
- Big business; petroleum industries, harbour associations
- NGO; nature protection organisations, bird watchers
- Chairman (leader of the negotiation in the role play)
- Local authorities
- Scientists representing universities.

Before the seminar, participants were issued with role instructions, which gave them suggestions about what positions and opinions a particular role should have, in order to help them act out. The roles were distributed between participants according to their background; for example a person working in local authority played the local authority role. Through this, participants were given the possibility to use their own experience from their daily work.

To prepare the course participants for the role play, a general storyline and negotiation instructions for each role were also distributed before the seminar. The general story included the following SD policy which participants had to discuss during the negotiation process:

- SD involves three main sectors, social, economic and environmental, and a balance must exist between these three sectors.
- SD is development that meets the needs of the present without compromising the ability of future generations to meet their own needs.
- Harbour structures should be planned from a SD perspective.
- In the harbour planning process, the factors that civilians in the country/countries believe to be the most important must be taken into consideration.

Photo 2: Enacting the role play during Seminar 2 in Stockholm

The first seminar day started with a brief introduction to familiarise the participants with the role play storyline and structure. The pre-negotiation part then

began. During this, there were bilateral discussions and participants had to find out as much as possible about each other's opinions on different topics and create alliances. In the formal negotiation phase, there were six topics on the agenda to be negotiated. Participants were expected to present their own ideas about the topics and to reach agreement/consensus about the topics; a two-thirds vote was required for approval. The role play finished with a presentation of the group results, the differences between the results were discussed and the participants shared their opinions about the role play as a whole.

3.2.1 The results of Seminar 2

3.2.1.1 Group 1

Group 1 agreed that the members of the PHC should be similar to the TC, all members should have the same voting rights and the chairman of the commission should be on a rotating basis between all TC members.

It was also agreed that the PHC guiding principles should be based on the harbour SD policy but some extra principles were added and underlined:

- The harbour should not be located in the drinking water supply or in the fishing areas.
- Recreation areas for future generations, the nature reserve and the bird island had to be protected.
- New energy sources had to be studied.

The PHC would make the final decision about harbour location and the governments of both states would ensure that the harbour was constructed according to the guiding principles and future operation.

To secure the financial resources, the costs of harbour infrastructure and ownership would be shared between both states and the commercial companies would pay for the superstructure. In the event of an accident, the polluter pays principle would be used and a fund established.

To establish broad support in the process of finding and deciding the new harbour location, the public would be involved by hearings and giving feedback. The documents of the PHC would be published on the internet before the decision about the harbour was made by the PHC and the PHC would take public needs into account.

Before the PHC could decide about the harbour location, an environmental impact assessment (EIA) would be needed and both governments and the businesses concerned would have to cover the costs of the EIA and further environmental controlling.

3.2.1.2 Group 2

Group 2 established restrictions concerning membership and chairing of the PHC. Members should be similar to the TC but with only one member representing small businesses and one member representing NGOs. The chairmanship would change on rotating basis between local authorities.

The PHC guiding principles would be based on the harbour SD policy. The PHC would be responsible for investigating the harbour location and the government would make the final decision. Public opinion would be important but would not be the only factor in the decision-making process.

The government and local authorities would have to secure and control the financial resources for the harbour construction and make a risk agreement to secure financial resources in the event of an accident.

A referendum about the harbour location would not be necessary for establishing broad support in deciding about the new harbour location because the documentation for the harbour decision-making process would be made public. The PHC would decide about the structure of the information campaign (why the harbour is needed and how it should be constructed) and which documents should be made public and when.

The location of the harbour with the smallest impact on the environment would have to be chosen or special measures would have to be introduced to minimise the impacts or to compensate for them. All investigations, for example EIA, would have to be carried out by specialists (scientists). The government and the local authorities would have to decide who would pay for these investigations.

3.3 Course evaluation

3.3.1 Questionnaire

To study whether the opinions of the course participants about SD and stakeholders changed during the course, a questionnaire was used during both seminars. In the questionnaire the participants listed which factors they thought were important in a harbour planning process. According to the results of this questionnaire, there was no obvious change in the opinions regarding SD.

3.3.2 Results from the course evaluation

To evaluate the course and to gather participants' experiences, especially about the role play, the participants were asked to answer an internet-based questionnaire on the course website. The questionnaire comprised 18 questions about the role play as a method, storyline, roles, chosen topics and the structure of the play. Most of the questions had fixed answers (yes absolutely, partly, good, hardly and not at all). For some issues more information was required and participants had to write their own answers. Twenty participants answered the questionnaire.

The majority of the course participants who responded thought that the role play as a whole was a success and that the discussions during the role play were meaningful. The storyline was considered relevant for studying conflict resolution and SD, but some of the participants thought that the role play storyline and the role characters should be described more clearly and be more detailed, especially if the role play is to be used in an undergraduate course. The participants gave good suggestions about new roles that could be added in the role play, for example society's role, an additional secretary and a chairman with experience of leading negotiation processes.

The general opinions of the participants on the course structure as a whole were very positive and the chosen topics were regarded as significant for the role play. Opinion about how well the role play mirrored a real conflict and provided participants with general knowledge and understanding of conflict resolution varied. Finally the role play as a whole was considered to be successful, the engagement of course participants during the play very good and the majority of the participants believed that role play is a useful method for learning about conflict resolution in coastal zone management.

4. Discussion and conclusions

Several interesting aspects emerged during the discussion about the concept of SD at the first seminar in Riga. The participants had no difficulties in defining the concept of SD as a common vision, but there were disputes regarding what kind of SD criteria should be used specifically in a harbour planning process.

The task for the five groups during Seminar 1 was to choose one harbour location from four alternatives given using their own SD criteria and conflict analyses. Despite the fact that the groups used a similar methodology, the results differed. Of the four possibilities given in the virtual case, all groups avoided alternative D (nature reserve), two groups chose alternative A, two alternative C and one alternative B. One group used 'worst case' as a start before they looked at different criteria.

The results from the role play during Seminar 2 were also interesting. Two groups enacted the same role play in parallel but reached different outcomes. The task was to decide the structure and tasks of the proposed permanent harbour commission (PHC). The results showed how different aspects and values can come into expression during a negotiation process, as shown in the group reports. For example:

The idea of democracy

Group 1 decided that all stakeholders who were members of the original commission should also be members of the permanent commission, and all of them should have the same voting rights and share the chairmanship. They gave the permanent commission the right to make the final decision. Group 1 also decided that all material should be made public and published on the internet. Group 2 excluded some of the stakeholders representing NGOs and smaller companies from the permanent commission and delegated the chairmanship to the local authorities and the decision-making rights to the government. They also thought that the permanent commission should decide which documents should be made public.

The vision about SD

Group 1 expressed an ambition to develop the SD harbour policy and concretised what SD meant for their region (virtual case study); in other words they tried to use the general vision of SD in a more local way. Group 2 adhered to the suggested general SD harbour policy given to them in the beginning of the role play.

Future accidents

There was also a distinct difference concerning the topic of who should pay in the event of an accident. Group 1 assumed the polluter pays principle, while Group 2 decided that the government, the local authorities and investors should take the economic responsibility for any future accidents.

Several interesting features about conflict resolution emerged during seminar discussions. When working with the task of locating a new harbour, the strengths and weaknesses of the SD concept/approach were visualised. The strengths were identified as the vision of SD and a feeling of working in the same direction, which does not create any greater conflicts. The weaknesses were visualised when trying to apply SD ideas in a tangible case, a local planning process. During the discussions, stakeholders' values affected the results regarding e.g. what interests should be prioritised, how local and regional conflicts were described and understood, etc.

The ambition of the course was to visualise differences (natural, cultural, etc.) between stakeholders, opinions about legal and market aspects, democracy, political systems and gender. Awareness of these differences was expected to have positive effects on people's understanding of each other and in that meaning contribute to conflict mitigation.

The exercise during Seminar 1, when the participants were forced to establish a common position about one of four suggested harbour locations, clearly showed participants that the decision-making process is not only based on facts and science but in fact on values, attitudes and the ability to handle these. This became even clearer during the role play, where the ideas of democracy, transparency and stakeholder participation in the decision-making process were reviewed. In our opinion, this shows how discussions and participation are important parts of the democratic process for SD and conflict resolution applied to coastal zone management of land and water.

We believe that the structure of the course and the pedagogic exercises, with one virtual case and one role play, increased participants' knowledge and awareness about conflict resolution and SD. It visualised different values and attitudes to central questions for SD and conflict resolution, such as democracy, publicity and stakeholders' possibilities to participate in and influence the planning process.

The specific results of the group discussions and negotiations in the role play and the results of the group work during Seminar 1 naturally reflect a number of important conditions linked to the course implementation, such as group composition as regards nationality, knowledge of languages, profession, gender, age and experience in this subject area. The course had a heterogeneous composition of participants reflecting all these different perspectives (companies, planners, genders, nationalities etc.), which is important in a course such as this, because it creates a dynamic process. The heterogeneous composition of groups is also the driving force in visualising different values and attitudes, which are often important reasons behind conflicts about the use of land and water in the coastal zones. This makes it possible to apply the course in areas with the focus on other conflicts.

5. References

Henriksen, Thomas (2004). On the Transmutation of Educational Role-Play, A Critical Reframing to the Role-Play on Order to Meet the Educational Demands (www.ropecon.fi/brap/ch10.pdf).

Chapter 5

Conclusions

Walter Leal[1], Nils Brandt[2], Dörte Krahn[1], Ronald Wennersten[2]

Many lessons have been learned throughout the duration of the COASTMAN project, some of which are described in the different sections of this Handbook.

The first lesson is that there is a need for deep insight into the subject of conflict resolution, in order to allow a holistic understanding of the various variables that influence resolution. There has to be general acknowledgement that conflicts occur whenever different interests collide. In the case of coastal zones, such interests often involve economic and sometimes political issues, which are played out against complex environmental backgrounds. In this regard, the project shows the strength of using 'sustainable development' (Brundtland) as a common framework for discussions. Using this framework creates a holistic view of the conflict resolution, while revealing differences in e.g. culture, legislation and attitude between countries around the Baltic Sea. Sustainable development can be a platform for formulation of a common vision, which can be a powerful tool in communication of potential outcomes to existing problems.

The second lesson to be learned from the project concerns the importance of stakeholder participation. In all case studies, it was essential to describe and analyse all stakeholder roles and interests, and to focus on the participation of the stakeholder when applying methods of conflict resolution. All the cases demonstrated that a true bottom-up perspective provides the most effective means of analysing conflicts and of avoiding long-term intractable conflicts.

A further lesson learned is that the process of understanding environmental conflicts and attempts to work towards conflict resolution may be greatly assisted by the use of tools. Such tools, whether theoretical (SWOT analysis) or practical (e.g. decision support) can lead to better conflict management and are potentially useful in conflict resolution or at least in holistic understanding of conflicts. In the case studies presented in this Handbook, emphasis was placed on various tools such as the Marker self-assessment tool, Frame analysis, hearings, interviews, SWOT analysis and the DPSIR Framework, which were applied to different cases.

Tools have the advantage of contributing to the decision-making process by improving its transparency, while at the same time ensuring that all interests are

1 TuTech Innovation GmbH – Hamburg – Germany.
2 Royal Institute of Technology (KTH) – Department of Industrial Ecology – Stockholm – Sweden.

taken into consideration. It should be noted that in order to succeed, the choice of tools has to ensure that due consideration is given to all different perceptions and concerns of the interested parties.

Another lesson that can be drawn from this Handbook and the case studies described therein is that holistic conflict understanding is essential in coastal zone management in order to ensure that the coastal environment and coastal ecosystems are conserved in the long-term. In particular, the following items were identified:

- In Hamburg (Germany), the usefulness of an efficient instrument to measure and communicate the environmental impacts of port activities was demonstrated.
- In Haapsalu (Estonia), a forum for scenario development and harmonisation of interests was established and stakeholders representing different interests were networked.
- In Klaipėda (Lithuania), the use of strategic environmental and risk assessment led to better understanding of the problem and compensation measures were identified.
- In Primorsk (Russia), ecological studies combined with social-ecological studies provided a solid basis for decision-making and assisted the local and regional authorities with future plans.
- In Stockholm (Sweden), developments relating to the Loudden docks area were assessed by means of various tools, which provided a holistic view of the problem.
- In Ventspils (Latvia), the large investments in environmental protection may make the transhipment of oil products and chemicals safer and hence reduce the likelihood of risks and conflicts associated with such transhipments.

The case studies also showed that there are no infallible recipes or 'silver bullets' that allow a given conflict to be solved in a pre-determined way, since conflicts vary in nature, degree of impact and potential environmental damage. Indeed, the case studies showed that:

1. Holistic understanding of a conflict and associated problems is needed in seeking solutions
2. Relevant stakeholders need to be identified
3. The positions of stakeholders and the pressures they face need to be understood.

These are important pre-conditions, without which no sustainable solutions can be found.

Some further noteworthy areas of action based on the needs identified in the Handbook include:

- Action on Coastal Areas vulnerable to hazards: Some coastal areas are already vulnerable to coastal hazards and the risks may be exacerbated by economic activities such as tourism or industry. Action in such areas is needed with a view to reducing the risks of such hazards being transformed into accidents
- Dredging and Spoil Disposal: This activity, which is vital in some coastal areas, must be carefully managed to avoid adverse effects on navigation, flood flow capacity and environmental quality
- Fisheries Management: Fishing, both recreational and commercial, has an impact on biodiversity and requires efforts to protect and improve stocks of fish species.
- Sustainable Port Activities: Despite the undisputed economic value of port activities such as the transport of goods, oil or chemicals, these activities need to be performed in such a way that the development and enhancement of coastal port infrastructure does not endanger the biodiversity of the region
- Public participation: More efforts are required to engage the public in activities such as boating, fishing, walking, picnicking, sightseeing and other recreational pursuits associated with coastal areas and to create opportunities for public participation in a variety of coastal issues

Finally, it is important to observe that in analysing issues relating to conflicts and in working towards conflict resolution, relevant data relating to native or non-native (invasive) aquatic or terrestrial plant and/or animal species are needed. This may allow a better preparedness to understand and deal with the conflict and its associated impacts and effects.

About the Authors

Haapsalu

Dr. *Arvo Iital* is an assistant professor at the Department of Environmental Engineering, Tallinn University of Technology. He has 13 year experience in monitoring and assessment of surface water quality, assessment of environmental state, eutrophication of inland and coastal waters and environmental policy. Currently he is teaching several courses at the university including environmental assessment, environmental policy and landscape ecology.

Contact: Arvo Iital, Tallinn University of Technology, Department of Environmental Engineering, Ehitajate tee 5, 19086 Tallinn, Estonia, e-mail: arvo.iital@ttu.ee

Krista Vilta, managing director of the Environmental Department, has been working for the of Haapsalu City Government for more than 8 years. Her key qualification is related to environmental protection, coastal zone management and environmental technology. She has been responsible for implementing many environmental and development projects in Haapsalu Municipality.

Contact: Krista Vilta, Haapsalu City Government, Posti 34, 90504 Haapsalu, Estonia, e-mail: Krista.Vilta@haapsalulv.ee

Enn Loigu, Prof., PhD, has more than 35 years research experience in assessment of surface water quality, eutrophication of inland and coastal waters, self-purification processes and water management. As a director of the Department of Environmental Engineering at Tallinn University of Technology he coordinates many research projects on national and international level.

Contact: Enn Loigu, Tallinn University of Technology, Department of Environmental Engineering, Ehitajate tee 5, 19086 Tallinn, Estonia, e-mail: enn.loigu@ttu.ee

Jüri Kurba has more than 30 years experience in the field of spatial planning and infrastructure management in Haapsalu town and in Tallinn. He has been involved in many national and international projects. At the moment he is in charge of the infrastructure compartment at the City Planning Department of Tallinn.

Contact: Jüri Kurba, Tallinn City Government, City Planning Department, Vabaduse väljak 7, 15198 Tallinn, Estonia, e-mail: jyri.kurba@tallinnlv.ee

Hamburg

Professor Dr. rer.nat., Dr. sci. habil., Dr. h.c. Walter Leal Filho is Head of Life Sciences at TuTech Innovation in Hamburg, and teaches in Hamburg, Stockholm and Naples, also being honorary professor at the A.I. Open University, Hochschule Zittau-Görlitz and the South-Western University "Neofit Rilsky"

(SWU). He has authored, edited or co-edited various books on the subject matters of innovation, has undertaken various innovation workshops and R&D training seminars in Central and Eastern Europe and has over 120 published papers to his credit. Professor Walter Leal Filho has coordinated the German elements of the Coastman project and organised a number of initiatives and has written a number of publications on the project.

Contact: Walter Leal Filho, TuTech Innovation GmbH, Life Sciences, Harburger Schlossstrasse 6-12, 21079 Hamburg, Germany, e-mail: leal@tutech.de

Agnieszka Holda and *Janne Juurikas*, both masters of science students in Environmental Engineering at the Technical University of Hamburg. They have worked within the Coastman project in the case study of Hamburg Harbour.

Irene Lucius, The Coastal Union Germany (EUCC), has scientifically investigated and implemented the ICZM-Marker application for TuTech Innovation GmbH in Hamburg within the Coastman project.

Dörte Krahn, Environmental Scientist, worked as a scientific assistant at TuTech Innovation GmbH, Germany. Her research interests are in environmental planning, marine, coastal, and aquatic ecosystems as well as in international policies of water issues. She worked as a project manager for the Coastman project and for other Interreg III B projects in the Baltic Sea Region.

Contact: Dörte Krahn, TuTech Innovation GmbH, Life Sciences, Harburger Schlossstrasse 6-12, 21079 Hamburg, Germany

Ahmed Quereshi, PhD student at the Technical University of Hamburg. He has undertaken a SWOT analysis for the Coastman project in the case study of Hamburg Harbour.

Klaipėda

Olga Belous, PhD, is lecturer in environmental management, environmental impact assessment and environmental law/ Klaipėda University, Klaipėda, Lithuania. Her research interests are in environmental education and environmental research. She integrates into scientific Committee of Natural Sciences and Mathematics Faculty.

Contact: Klaipėda University, Herkaus Manto 84, LT-92294 Klaipeda, Lithuania, e-mail: obelous@gmf.ku.lt

Saulius Gulbinskas, PhD, is lecturer in integrated coastal zone management, environmental impact assessment, Klaipėda University, Klaipėda, Lithuania. His research interests are in environmental education and environmental sedimentology, geology and coastal zone management. He integrates into Coastal Research and Planning Institute (CORPI) activities coordination as Vice-Director.

Contact: Klaipėda University, Herkaus Manto 84, LT-92294 Klaipeda, Lithuania

Primorsk

Elena Viktorova, MBA, project coordinator from Russian side, project management, interest in environmental education and research, conflict resolution education, integrated management of the territory.

Contact: Elena Viktorova, Saint-Petersburg State Polytechnical University, 29, Polytechnicheskaya str., 195251 Saint Petersburg, Russia, e-mail: Elena.Viktorova@mail.ru

Shilin Michael, Dr., Professor of Geography at the Russian State Hydrometeorological University and State Polytechnical University in St. Petersburg, Russia. His research interests are in ecological anthropology, coastal zone ecology and inter-cultural contacts in the coastal zone. He is editor of the Annual Bulletin of the Baltic Floating University, supported by UNESCO and International Oceanographic Commission.

Contact: Shilin Michael, State Polytechnical University, 195251 St. Petersburg, Polytechnical str. 29, Russia, email: Shilin@rshu.ru

Nikolay Plink, Head of Department on Integrated Coastal Management of the Russian State Hydrometeorological University (RSHU), Doctor of Science, Head of UNESCO/UNITWIN Network Cheir on „Wise Coastal practices and Sustainable Human Development – WiCoP-Europe" (RSHU).

Contact: Russian State Hydrometeorological University.

G.G. Gogoberidze

Contact: G.G. Gogoberidze, Russian State Hydrometeorological University

A.G. Kosheleva

Contact: A.G. Kosheleva, Local Administration of Primorsk City Municipality,

Stockholm

Professor *Ronald Wennersten* is head of the Department of Industrial Ecology at Royal Institute of Technology (KTH) in Stockholm, Sweden. He got his PhD in Chemical Engineering at Lund University in 1981. After working in Industry he came to KTH in 1996. After he became head of the Department of Industrial Ecology in 2000, his ambition has been to merge the existing research in the areas of Environmental Management and Environmental System Analysis with his own research in Risk Management within the framework of Industrial Ecology and Sustainable Development. He is head of the Joint Research Centre for Industrial Ecology at Shandong University in China where he also is guest professor.

Contact: Ronald Wennersten, Royal Institute of Technology (KTH), Department of Industrial Ecology, Teknikringen 34, SE-100 44 Stockholm, Sweden, e-mail: Rw@ket.kth.se

Nils Brandt, associate professor, and director of studies at the Department of Industrial Ecology at Royal Institute of Technology (KTH) in Stockholm, Sweden. He is program manager for Sustainable Technology, an international masters program at KTH. His research interests include climate strategies and action on local and regional level, and coastal zone management.

Contact: Nils Brandt, Industrial Ecology, School of Energy and Environmental Technology, Royal Institute of Technology, Stockholm (KTH), 100 44 Stockholm, Sweden, email: nilsb@kth.se

Jan Fidler, PhD, researcher with interests in risk methodology, risk communication, information analysis and environmental education. He is also lecturer in risk management courses at Royal Institute of Technology (KTH), Stockholm, Sweden.

Contact: Jan Fidler, Kungliga Tekniska Högskolan (KTH), School of Energy and Environmental Technology, Industrial Ecology, Teknikringen 34, 114 28 Stockholm, Sweden, email: fidler@kth.se

Åsa Larsson, masters of science student in Chemical Engineering at KTH. She has done her thesis work in the Coastman project in the case study of Loudden.

Contact: Åsa Larsson, Industrial Ecology, School of Energy and Environmental Technology, Royal Institute of Technology, Stockholm (KTH), 100 44 Stockholm, email: asalar09@kth.se

Ventspils

Ilga Zīlniece – head of Environmental Department in Ventspils City Council. She is responsible for different environmental issues in the municipality as well as cooperation with state environmental institutions.

Contact: Ilga Zīlniece, Ventspils City Council, Head of Environmental Department, Andreja iela 3, Ventspils, Latvia, email: ilga.zilniece@ventspils.gov.lv

Tatjana Valdmane – chief specialist of Environmental Department in Ventspils City Council. She is responsible for creating environmental licenses for different enterprises and safety issues in potentially dangerous enterprises.

Contact: Tatjana Valdmane, Ventspils City Council, Chief Specialist of Environmental Department

Ineta Kraule – specialist of Environmental Department in Ventspils City Council. She is responsible for protection of natural environment an Blue Flag movement.

Contact: Ineta Kraule, Ventspils City Council, Deputy Head of Management and Information Department

Sofija Galindoma – specialist of Organizing and Information Department in Ventspils City Council, interpreter in the COASTMAN project.

Contact: Sofija Galindoma, Ventspils City Council, Deputy Head of Management and Information Department

Laila Kule – lecturer in Latvian University Faculty of Geography and Earth Sciences.

Contact: Laila Kule, University of Latvia

Thematic Index

Further reading

1. *Cicin-Sain, B. & Knecht, R.W.* (1998). Integrated Coastal and Ocean Management: Concepts and Practices. Washington, DC, Island Press.
2. *Clark, J.R.* (1996). Coastal Zone Management Handbook. Boca Raton, CRC Press.
3. *Clark, J.R.* (1998). Coastal Seas: The Conservation Challenge. Oxford, Blackwell.
4. *French, P.W.* (1997). Coastal and Estuarine Management. London, Routledge.
5. *Goldberg, E.B.* (1994). Coastal Zone Space – Prelude to Conflict? Paris, UNESCO.
6. *Hinrichsen, D.* (1998). Coastal Waters of the World. Washington, DC, Island Press.
7. *Kay, R.C. & Alder, J.* (1999). Coastal Planning and Management. London, Routledge.
8. *Viles, H. & Spencer, T.* (1995). Coastal Problems: Geomorphology, Ecology and Society at the Coast. London, Edward Arnold.
9. *Korakandy, R.* (2005). Coastal Zone Management : A Study of the Political Economy of Sustainable Development. Kalpaz, Delhi.
10. *Shipman, B. & Stojanovic, T.* (2007). Facts, Fictions, and Failures of Integrated Coastal Zone Management in Europe. In Journal of Coastal Management, Volume 35, Issue 2 & 3 April 2007, pp. 375-398.

Umweltbildung, Umweltkommunikation und Nachhaltigkeit
Environmental Education, Communication and Sustainability

Herausgegeben von/Edited by Walter Leal Filho

Band 1 Walter Leal Filho (Hrsg.): Umweltschutz und Nachhaltigkeit an Hochschulen. Konzepte-Umsetzung. 1998.

Band 2 Walter Leal Filho / Farrukh Tahir (Eds.): Distance Education and Environmental Education. 1998.

Band 3 Walter Leal Filho (ed.): Environmental Engineering: International Perspectives. 1998.

Band 4 Walter Leal Filho / Mauri Ahlberg (Eds.): Environmental Education for Sustainability: Good Environment, Good Life. 1998.

Band 5 Walter Leal Filho (ed.): Sustainability and University Life. 1999. 2nd, revised ed. 2000.

Band 6 Wout van den Bor / Peter Holen / Arjen Wals / Walter Leal Filho (eds.): Integration Concepts of Sustainability into Education for Agriculture and Rural Development. 2000.

Band 7 Manfred Oepen / Winfried Hamacher (Eds.): Communicating the Environment. Environmental Communication for Sustainable Development. 2000.

Band 8 Walter Leal Filho (ed.): Communicating Sustainability. 2000.

Band 9 Walter Leal Filho (ed.): Environmental Careers, Environmental Employment and Environmental Training. International Approaches and Contexts. 2001.

Band 10 Rolf Jucker: Our Common Illiteracy. Education as if the Earth and People Mattered. 2002.

Band 11 Walter Leal Filho (ed.): Teaching Sustainability at Universities. Towards curriculum greening. 2002.

Band 12 Walter Leal Filho (ed.): International Experiences on Sustainability. 2002.

Band 13 Lars Wohlers (Hrsg.): Methoden informeller Umweltbildung. 2003.

Band 14 Ulisses Azeiteiro / Fernando Conçalves / Walter Leal Filho / Fernando Morgado / Mário Pereira (eds.): World Trends in Environmental Education. 2004.

Band 15 Walter Leal Filho / Arnolds Ubelis (eds.): Integrative approaches towards sustainability in the Baltic Sea Region. 2004.

Band 16 Walter Leal Filho / Michael Littledyke (eds.): International Perspectives in Environmental Education. 2004.

Band 17 Daniela Krumland: Beitrag der Medien zum politischen Erfolg. Forstwirtschaft und Naturschutz im Politikfeld Wald. 2004.

Band 18 Walter Leal Filho / Bernd Delakowitz (Hrsg.): Umweltmanagement an Hochschulen: Nachhaltigkeitsperspektiven. 2005.

Band 19 Walter Leal Filho / Volker Lüderitz / Gunther Geller (Hrsg.): Perspektiven der Ingenieurökologie in Forschung, Lehre und Praxis. 2006.

Band 20 Walter Leal Filho (ed.): Handbook of Sustainability Research. 2005.

Band 21 Walter Leal Filho / Dieter Greif / Bernd Delakowitz (eds.): Sustainable Chemistry and Biotechnology – A Contribution to Rivers Management. 2006.

Band 22 Walter Leal Filho / David Carpenter (eds.): Sustainability in the Australasian University Context. 2006.

Band 23 Walter Leal Filho / Arnolds Ubelis / Dina Berzina (eds.): Sustainable Development in the Baltic and Beyond. 2006.

Band 24 Walter Leal Filho (ed.): Innovation, Education and Communication for Sustainable Development. 2006.

Band 25 Walter Leal Filho / Mario Salomone (eds.): Innovative Approaches to Education for Sustainable Development. 2006.

Band 26 Walter Leal Filho / Franziska Mannke / Philipp Schmidt-Thomé (eds.): Information, Communication and Education on Climate Change – European Perspectives. 2007.

Band 27 Ulisses Miranda Azeiteiro, Fernando Gonçalves, Ruth Pereira, Mário Jorge Pereira, Walter Leal Filho, Fernando Morgado (eds.): Science and Environmental Education. Towards the Integration of Science Education, Experimental Science Activities and Environmental Education. 2008.

Band 28 Walter Leal Filho / Nils Brandt / Dörte Krahn / Ronald Wennersten (eds.): Conflict Resolution in Coastal Zone Management. 2008.

Walter Leal Filho / Mario Salomone (eds.)

Innovative Approaches to Education for Sustainable Development

Contents: Guidelines of the Toronto World Congress for wide-ranging ...

Homepage http://www.peterlang.de